To Gladys.
    Xmas '85'

From -
    Tris & Vera.

THE AUDUBON SOCIETY

Les Line, Editor of Audubon magazine, and Walter Henricks Hodge

A Chronicle Press Edition

BOOK OF WILDFLOWERS

HARRY N. ABRAMS, INC., PUBLISHERS, NEW YORK

In memory of . . .
Hal Borland, who encouraged us to look beyond our doorsteps,
and Kenneth Hoffman, who looked and found the lady's-slippers
Les Line

For Bobbie, Wendy and Katsuko—the girls in my life
W.H.H.

Library of Congress Catalog Card Number: 78-6204
Library of Congress Cataloging in Publication Data
Line, Les.
The Audubon Society Book of Wildflowers.
1. Wildflowers—Pictorial works.   2. Botany—
Ecology.   I.   Hodge, Walter H., 1912—   , joint author.
II.   National Audubon Society.   III.   Title.
QK98 L75   582.13   78-6204
ISBN 0-8109-0671-6

Prepared and produced by Chanticleer Press, Inc.
Color reproductions by Nievergelt Repro AG, Zurich, Switzerland
Printed and bound by Amilcare Pizzi, S.p.A., Milan, Italy

Chanticleer Staff:
Publisher: Paul Steiner
Editor-in-Chief: Milton Rugoff
Managing Editor: Gudrun Buettner
Project Editor: Susan Rayfield
Assistant Editor: Richard Christopher
Production: Patricia Pascale, Ray Patient
Design: Massimo Vignelli

*Note on Illustration Numbers: All illustrations are numbered
according to the pages on which they appear.*

First frontispiece: *A carpet of California wildflowers—
lupine, poppies and owl's clover* (Carl Kurtz)
Second frontispiece: *A field of California poppies*
(Eschscholtzia californica) *in the Mojave Desert* (Carl Kurtz)
Third frontispiece: *California poppies* (Carl Kurtz)
Fourth frontispiece: *California poppies* (Sally Myers)
Fifth frontispiece: *A California poppy* (Ed Cooper)

Contents

In a more gentle time, when children grew up in small country towns rather than sprawling suburbs, the first day of May meant something very special. May Day. The image it evokes today is one of awesome armaments being paraded through Moscow's Red Square under the threatening gaze of the Russian hierarchy. But when I was a boy, in a place called Sparta, Michigan, it meant a knock on the door and a quick dash to a hiding place from which one could watch a mother or a girl-next-door or a nice lady across the street being pleasantly surprised.

May Day. Spring had arrived, by the calendar, a week or so earlier. In the woodlots of beech, maple, oak, and hickory that had been preserved by the farmers who cleared this rolling land a century before, spring wildflowers were blooming in wondrous profusion: bloodroot, trout lilies, trilliums, hepatica, violets, anemone, trailing arbutus, Jack-in-the-pulpit, Dutchman's breeches, marsh marigold. As the turn of the month approached, we secretly made neat little flower baskets out of brightly colored construction paper, cut out handles and glued them on with flour paste. And before breakfast on May 1st—very early indeed if it was a school day—we would bicycle to the nearest woodlot, pick our bouquets, rush back, and hang them on the appropriate doorknobs.

The woodlot nearest to my home was Twin Hills—a double bump on a rutted country lane a mile west of town. It was the first woods I ever knew. On spring Sundays after church, my father would drive the family there in his sleek black Chevrolet to hunt for morel mushrooms and to pick wildflowers for the dinner table. I always managed to get my best shoes and pants muddy in the wet places where Jacks-in-the-pulpit grew. Who owned these woods I never knew nor cared. No fences, no signs thwarted entry, and I always visited them freely.

My father's death brought me home this past April. On a beautiful spring afternoon, after family and friends had departed, I drove to Twin Hills, expecting to find the woods replaced with split-level homes. Indeed, two or three houses had been built amidst the trees. But otherwise the woods were much as I remembered. The beeches and maples were larger, of course, though they had seemed very large to a small boy. And the spring flowers were everywhere. I had no May baskets to fill, no table to decorate. I had a dinner engagement and did not want to muddy my best shoes and pants. So, from the car window, I admired a close-by clump of bloodroot, their impeccably white flowers backlit by the late sun. And I drove on, wondering if children ever hang spring bouquets on doorknobs anymore.

The legacy of those boyhood trips to the spring woods was a lifetime interest in wildflowers. And no opportunity to do a book has excited me more than this one, the third in a series brilliantly conceived by Paul Steiner of Chanticleer Press, companion to *The Audubon Society Book of Wild Birds* and *The Audubon Society Book of Wild Animals*. It is a basket of fresh wildflowers presented to the world, a spectacular bouquet of 181 photographs assembled by picture editor Susan Rayfield to represent every major habitat on Earth—from the deciduous woodlands where I first hiked to the treeless barrens of the Far North. Each chapter is introduced by an authoritative, informative essay by my coauthor, Walter H. Hodge, who has "botanized" the world around. As in the previous books in this series, my captions accompany the photographs, telling the reader of the natural history, lore, and etymology of each flower chosen. I shall consider our efforts a success if, for a few readers who in their childhoods were deprived of firsthand knowledge of the wonder of wildflowers, they awaken an interest as passionate as my own.

Les Line

14. *In the rain forest of Trinidad's low mountains, epiphytes occupy every available niche on horizontal tree limbs. Among these air plants are ferns and orchids, bromeliads, and even cacti. They are not parasites on their host tree; rather, epiphytes have adopted a life high in the air in order to be nearer the sunlight and have evolved special ways of gathering and storing water and nutrients.* (Erwin A. Bauer)

Clingers and Climbers

The massive spreading limbs of an old Trinidad rain tree (*Samanea*) are decorated with an unbelievable number of air plants, uninvited guests thriving in this soil-less aerial habitat. Among this company of plant aerialists are numerous creeping ferns and peperomias, bromeliads, many orchids, and the strange mistletoe cactus (*Rhipsalis*). Scores of different plant species may grow upon such a host tree in the wet tropics, including herbs, shrubs and even small trees. In 1492, five days after his epic landfall, Columbus recorded seeing strange trees with "branches of different kinds, all on one trunk; one branch is of one kind and one of another, and so unlike that it is the greatest wonder in the world. . . ." Unknowingly he was reporting a common sight in the tropics, a tree full of air plants.

Botanists call air plants epiphytes. They are characteristic of wet tropical forests where the year-round humidity permits one of the most unusual of plant life-styles. So abundant are they that the uninitiated visitor assumes them to be parasites obtaining food and water from the living tissues of their host. On the contrary, the air plants of the rain forest merely sit on their hosts, high in the air, so as to reach the sunlight needed by all green plants for synthesizing food. Without access to the soil and its important sources of water and minerals, epiphytes are faced with serious problems of survival. They have solved these by evolving special adaptations for storing water and nutrients. Air plants cannot live in temperate forests, where daily rains are lacking. Even worse is the prolonged cold and drought of winter when water, frozen into ice or snow, is unavailable for months. Thus the forests of temperate lands have few epiphytes. Those that occur are nonflowering kinds—algae, lichens, mosses and perhaps an occasional fern. As one moves toward the equator, winters are shorter and milder and eventually disappear. At the same time some of the more hardy epiphytes of the tropics begin to appear. These are mostly diminutive orchids or, in the western hemisphere, species of bromeliads. The familiar Spanish moss (*Tillandsia usneoides*), a bromeliad rather than a moss, is a tropical wildflower that can thrive epiphytically in mild, warm temperate forests. There it occupies treetop sites, free of the savage competition for space it meets farther south.

In tropical rain forests, whether lowland or montane, air plants are everywhere. The living ladders of the forest—giant boles and tree branches, contorted cables of the giant lianas and plummeting roots of arborescent epiphytes—are the seats on which air plants perch. And in some cases, tiny orchids cling to the surface of single host leaves. Because of the variations of light and humidity that occur from the ground to the canopy, each level of these living ladders supports a different epiphytic company. Close to the ground, where humidity is highest and light level the least, flowering epiphytes are few. Liverworts, mosses and delicate filmy ferns are among plants that can thrive in the deep twilight and high humidity near the forest floor. Higher up, at the halfway level, light has so increased as to permit a new epiphytic flora to appear. Mid-level air plants are more drought-resistant because humidity there is less. Ferns with leaves less susceptible to drying replace the more delicate "filmies," and flowering epiphytes are then more common. The latter are mostly shade-lovers—creeping peperomias, colorful gesneriads, the bromeliads and aroids, and even climbing cacti. Many of these air plants have succulent stems or leaves that serve to store water for use in the rainless periods that occur even in a rain forest.

As one climbs skyward, the mid-level epiphytes are replaced by another company of wildflowers that cling to the highest rungs of the forest ladder. Hidden there are many of the showier orchids and bromeliads, herbs and even trees. Some of the latter may develop into stranglers that ultimately envelop and destroy their host. Many of this treetop company are xerophytes, plants with leaves specially developed to endure the

ing effect of full sun and occasional wind. To
accomplish this a variety of moisture-storing devices
have evolved.

Orchids and bromeliads of these hanging forest gardens
have especially good protection. Many cover their
moisture-laden tissues with a waxy skin, nature's
waterproof wrapper, which effectively retards evapora-
tion. Others, particularly aroids and orchids, form
swollen stems (pseudobulbs) to better store emergency
supplies of water for "non-rainy" days. Some orchids
may even shed their leaves like deciduous trees, if a
dry period persists. Bromeliads have silvery microscopic scales on their leaf surfaces which are able to
absorb water directly from the air. Rosettes of erect
overlapping leaves that form watertight tanks or
"rain barrels" are a familiar structure among many
bromeliads. Some of these resemble elegant vases
which, when the plant is in blossom, appear to be dis-
playing their own beautiful flower arrangements.
Others, sometimes three feet in diameter, are more
like great tubs. Enough water may accumulate in these
receptacles to last through long rainless periods.
These natural aerial aquaria invariably support tiny
aquatic communities, including algae, frogs, salamanders
and various invertebrate creatures. Debris inevitably
accumulates in such tanks, yielding abundant mineral
nutrients for the air plant.
Other epiphytes, especially the giant Malayan queen-
of-orchids (*Grammatophyllum speciosum*), as well as
the American tail flowers (*Anthurium*), form similar-
appearing structures with their large leaves. These do
not hold water, however, but serve rather as giant
wastebaskets to catch leaves or other falling debris; the
subsequent decomposition produces humus, which is
soon penetrated by special feeder roots seeking the
mineral nutrients it contains. One of the strangest
methods of obtaining nutrients is shown by certain
epiphytes, chiefly Indonesian species of the madder and
milkweed families (Rubiaceae and Asclepiadaceae),
that have developed interesting symbiotic relationships
with ants. Colonies of ants live within the swollen stems
or leaves of these plants in cavities seemingly evolved
for this unusual purpose. In return for shelter, the ants
not only protect the plant from other insects but also
provide it with the offal of their activities. Such
nitrogenous material is sought out by the epiphytes'
specialized feeder roots.

The tie-roots of many epiphytic orchids and aroids are covered with a whitish absorbent material called velamen, which can soak up water easily from rain, mist or fog. Under the velamen lies a green photosynthetic tissue permitting orchid roots to manufacture food just as leaves do. Such photosynthetic roots attain their ultimate expression in the curious tapeworm orchids (*Taeniophyllum*) of the Asiatic tropics. These tiny epiphytes lack leaves completely, and so the orchid must rely entirely on its green, tapeworm-shaped roots for all food manufacture.

The cloud forest is a botanist's paradise, for here the wet tropical forest puts on its best display of wildflowers. In the New World tropics, bromeliads are always prominent. Notable wildflowers of Indomalaysia include many little-known rhododendrons (*Rhododendron*)—all epiphytes—as well as numerous tropical pitcher plants (*Nepenthes*), which thrive in the mossy forest's treetop bogs. But wherever wet montane rain forests occur, orchids are among the most numerous epiphytic wildflowers.

The worldwide assemblage of orchids, numbering some 25,000 species, is the largest of all plant families. Its members range in size from pygmies less than an inch high to the ten-foot-tall queen-of-orchids of Malaya. Most orchids are air plants of the wet tropics, yet some species are to be found in each of the world's major habitats. A look at their relative numbers in each gives an idea of the luxuriance of the wet tropics as the world's optimal habitat. The kinds of tundra orchids can be counted on one hand; those of the taiga may number a dozen; in the forests, grasslands and deserts of temperate North America there are about 200; but in the wet tropics the kinds of orchids are legion. Thus in the wet montane forests of Costa Rica, the orchids, mostly epiphytes, number 900 species, or four times those of Canada and the United States combined. It is appropriate that an epiphytic orchid, the beautiful guaria morada (*Cattleya skinneri*), is the national flower of Costa Rica.

19. *A ghost orchid (Polyrrhiza lindenii) glows in the dim rece⸺ of Corkscrew Swamp, a Nation⸺ Audubon Society sanctuary in s⸺ Florida famous for its virgin sta⸺ of bald cypress trees. Found only⸺ in that subtropical state and Cuba, this orchid suggests, one naturalist wrote, "a giant spider clutching the trunks of trees." Several of the fragrant, greenish-white flowers grow on each slender stem.* (M. Philip Kahl)

20 *overleaf. These stunning orchids—stunning in shape and color—belong to the genus* Masdevallia, *with some 300 species in the tropical American highlands. The showy sepals unite at the base to form a tube, which ends in a long tail. Many epiphytic orchids are linked to their host trees with tie roots, which sponge water from rain or mist and contain photosynthetic tissue that manufactures food as a leaf would.* (M. P. L. Fogden)

22. Cattleya bowringiana *beautifully represents a familiar group of orchids, for flowers of this genus are the ones most often grown by hobbyists and worn on special occasions as corsages. Hundreds of varieties have been developed from the forty species that occur naturally in tropical America. Cattleya orchids have been adopted as the national flower of Colombia, Costa Rica, and Panama. Their fruit takes a year or more to mature and contains several hundred thousand microscopic seeds.* (James H. Carmichael, Jr.)

23 *Wax torch* (Aechmea bromeliifolia) *ranges from Belize and Guatemala to Argentina. Its generic name is derived from the Greek word for "point," referring to the rigid points on its floral envelope. The greenish-yellow flowers turn black as they mature.* (James H. Carmichael, Jr.)

24 *and* **26** *first and second overleaf. Cup of flame is an appropriate name for bromeliads of the genus* Neoregelia, *whose fifty-two species occur mostly in Brazil. The inner rosette of brightly colored leaves lures hummingbirds to pollinate the small, violet-blue flowers. With one exception, the 1,500 species in the bromeliad family are native to the western hemisphere.* (E. S. Ross; Erwin A. Bauer)

23

25. *A blooming period of several months characterizes orchids of the genus* Renanthera, *with fifteen species occurring throughout Southeast Asia and offshore into the Solomon Islands. The red dancing orchid (*Renanthera imschootiana*) is native to northeastern India.* (Stanley Breeden)

28. *Bright red flowers, like those of* Columnea microphylla, *are a clue that plants count on hummingbirds for pollination. Flowers in this New World genus produce such copious nectar that they are sometimes called* liana de syrupe. *(Katrina Thomas /Photo Researchers, Inc.)*

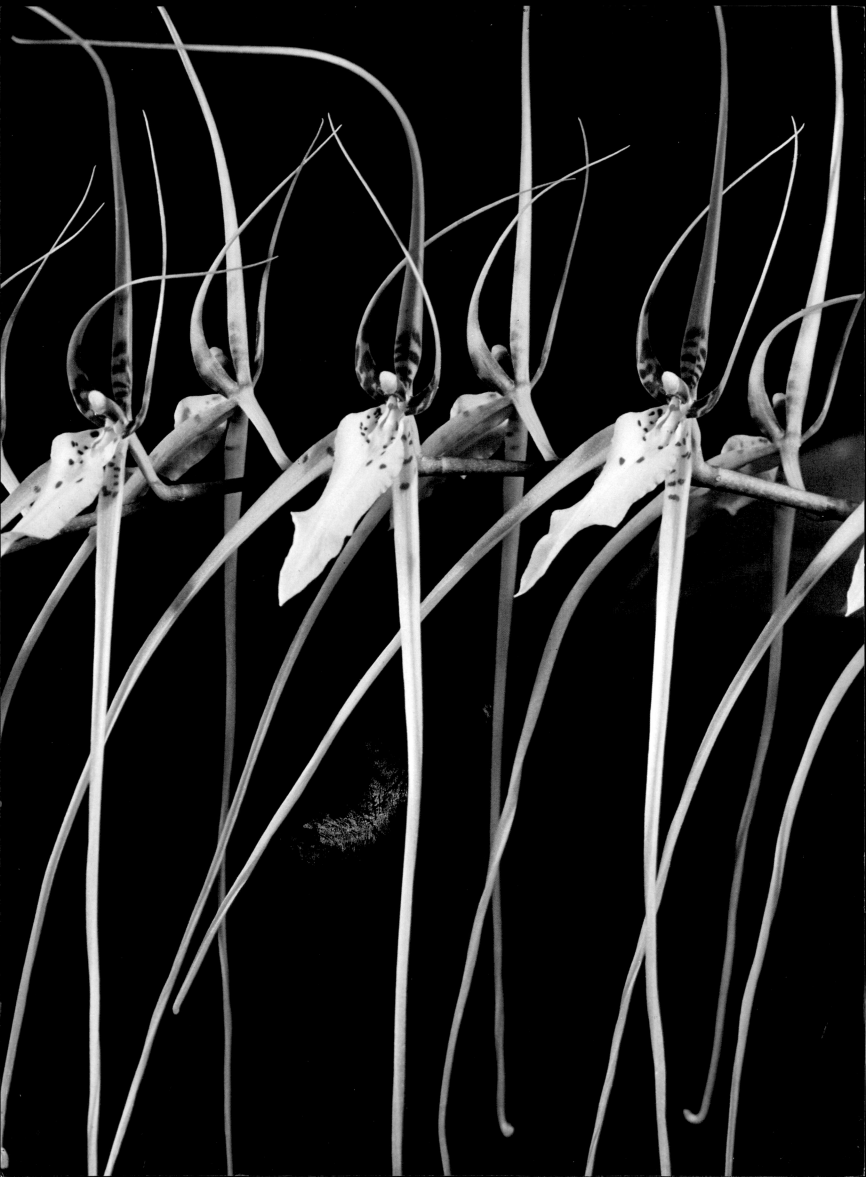

30. *A spider orchid* (Brassia longissima), *photographed at 6,000 feet in the Costa Rica cloud forest, bears a dozen flowers on a two-foot-long inflorescence. The tail-like sepals are twelve inches long. In one of the quirks of botanical nomenclature, this New World genus, which includes fifty species found in Central America and northern South America, was named for William Brass, who collected plants in West Africa(!) in 1782.* (Walter H. Hodge/Peter Arnold, Inc.)

31. *Queen of the night is a popular name for this night-blooming epiphytic cactus* (Epiphyllum oxypetalum). *Growing naturally from Mexico to Brazil, it is a giant among its kind, standing six feet tall with fragrant flowers ten inches wide. There are some twenty wild species of "orchid cacti" in this genus, from which horticulturists have created more than 3,000 spectacularly colored hybrids.* (Walter H. Hodge/Peter Arnold, Inc.)

33. Schomburgkia beysiana *is one of fifteen species of cowhorn orchids that grow from Mexico to Brazil. The genus is named for Richard Schomburgk, who explored tropical American forests in the nineteenth century.* (Kjell B. Sandved)

32. *The 400 species of epiphytic orchids in the genus* Oncidium, *found from Costa Rica to Ecuador, are called dancing-lady orchids. In most species the flowers appear at once in a spectacular spray. Among the exceptions is* Oncidium kramerianum; *its twenty buds open one at a time.* (James H. Carmichael, Jr.)

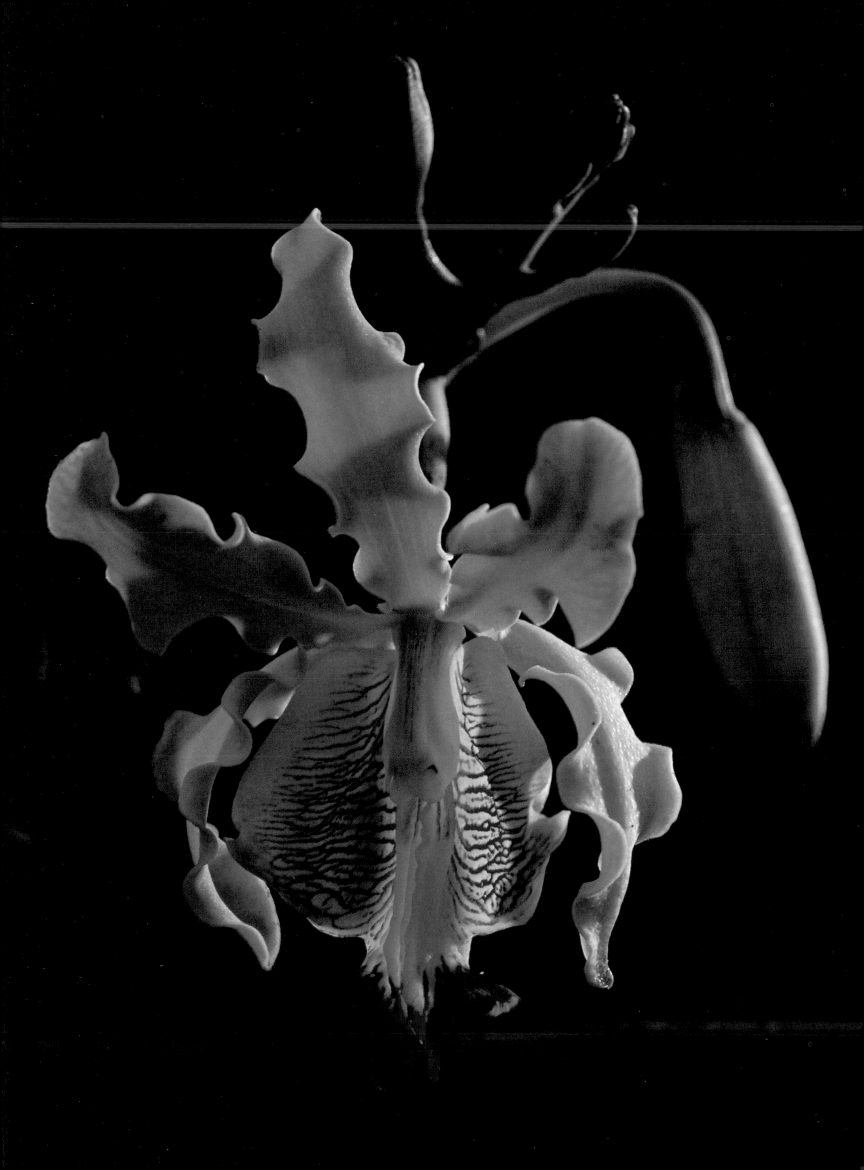

34. *More than a hundred beautifully mottled orchids, each three inches long, are borne on the branching inflorescence of* Renanthera storei, *native to the Philippines. Most of the world's 25,000 species of orchids —largest of all plant families—are tropical epiphytes. Some 900 species occur in the cloud forests of Costa Rica alone.* (Kjell B. Sandved)

35. *Fragrant flowers less than a quarter-inch across are evenly spaced on the fifteen-inch hanging clusters of golden chain orchid* (Dendrochilum filiforme)*. The name of this genus, which has 100 members from Southeast Asia to the Philippines, means "forest-loving."* (Kjell B. Sandved)

36 *overleaf. The bromeliad known as summer torch* (Billbergia pyramidalis) *has three-foot-long, vase-shaped leaves that hold a quart or more of water. This supply sustains the plant over rainless periods, and also provides a place for frogs, salamanders, snails, and insects to live and breed. The orange-red petals, tipped with violet, are characteristic of the fifty-two species in this genus. Birds eat their oily fruit and regurgitate the seeds, thus propagating the species. Summer torch is a popular, easily grown flower in tropical gardens.* (James H. Carmichael, Jr.)

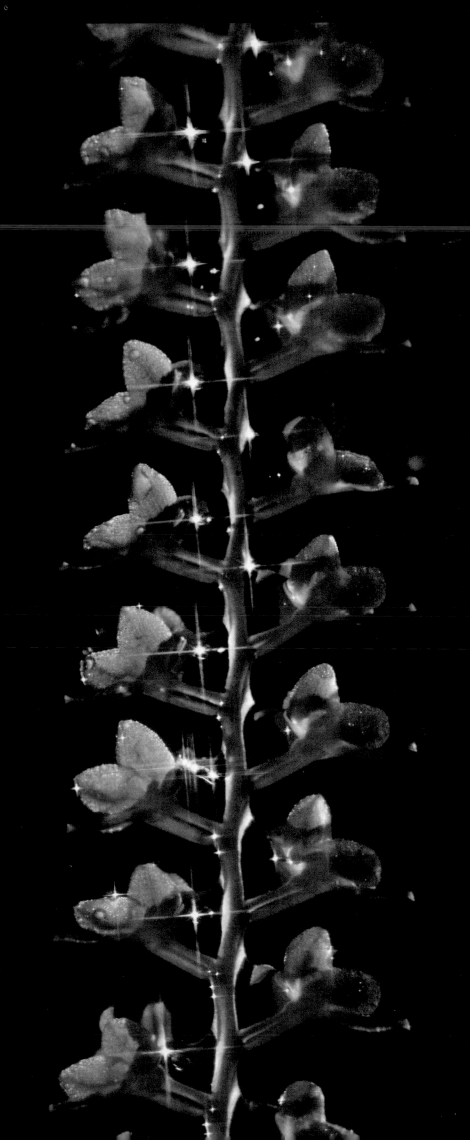

38. *The patterned foliage and intensely colored flower of flame violet (Episcia sp.) in the cloud forest of Venezuela testify to its kinship with such popular houseplants as African violet and gloxinia. Thriving in moist shade, flame violets create beautiful mosaics with their attractive, paired leaves, which are broad and thin to fully utilize the dim light. It is one of 1,200 species in the gesneriad family, found in greatest abundance in the natural greenhouse of the wet tropics, where more than 200 inches of rain may fall in a year and the temperature rarely slips below 80°F.* (Karl Weidmann)

Flowering Jungle

"A great and dark sea" is how the rain forest of the
New World tropics impressed Gonzalo Fernandez de
Oviedo in 1526. He added: "I say that the trees of these
Indies are a thing that cannot be explained, for their
multitude; and the earth is so covered with them in
many parts one cannot see the sky from below, and in
this respect one could say that this is a great and dark
sea; because though part is seen, much more is not."
The Spanish chronicler was alluding to a forest the
likes of which no European had even seen before—
a far cry from the sunny pine forests and open maquis
of his native Spain.

No one can stand in a rain forest without emotion.
Some 400 years after Oviedo, I traveled to one of the
Caribées to study rain forest plants. A woodcutter's
trail led me easily into the primeval depths. In such a
forest, with hundreds of arborescent species, almost
every tree is different. All are broad-leaved evergreen
hardwoods. Boles ten feet in diameter are common
and may rise 200 feet in the air, often from bases that
flare into curious massive buttresses. The foliage of
the canopy is so high above one's head that it defies
recognition. From the heights drop the cord-like roots
of arborescent epiphytes; numerous vines lift festooning
cables to the sunlight; while masses of aroids, ferns and
other air plants clutch at all available footholds. At
ground level light is dim, and the terrestrial vegetation
is sparse. The floor of a rain forest is not a jungle of
impenetrable vegetation. No wind reaches the ground,
and the stillness produces a solemn cathedral-like
setting. I was able to wander through that Caribee

forest almost as easily as through a New England woodland.

This is a fair description of any of the great rain forest areas of the globe, whether of the Amazon basin, the world's most extensive forest; of the Congo in Africa; or of the innumerable tracts of rain forest widely scattered through southeastern Asia to Indonesia, the Philippines and other archipelagos of the South Pacific. All these tropical forests look alike, but the component species are completely different, for unlike the taiga and deciduous forest of temperate lands, the world's rain forests have never been interconnected.

Wherever the lowland rain forest impinges on mountains, its overall appearance and the composition of its flora change. With increasing altitude, the dominating trees diminish in stature and the several layers of smaller understory trees, so characteristic of the optimal lowland forest, disappear. At cloud level in the mountains, fog is constant and precipitation is high. This is the region of the so-called cloud forest or mossy forest. For much of the year the vegetation is saturated with moisture, making it possible for mosses, ferns and flowering epiphytes to populate all levels of the habitat. As the timberline is approached, at about 10,000 feet, the forest decreases further in height and is often transformed into almost impenetrable thickets, interlaced with feathery tree ferns and bamboos. This pygmy rain forest is appropriately known as elfin forest.

The kinds of life, both animal and plant, which inhabit the lowland tropical rain forest number more than all the rest of the world's terrestrial biota together. The climate is optimal especially for plants. Equatorial sunlight not only is constant but shines an unvarying eight hours per day; the mean temperature is 80°F (27°C), while annual precipitation, often 200 inches or more, is abundant and, more important, well distributed throughout the year. Thus plant growth has been continuous for eons. Under such conditions the number of organisms has burgeoned, and their interrelationships, though still largely unknown, are more complex than anywhere else.

On the floor of a lowland rain forest, there is poverty among riches. For although the rest of the forest is incomparably diverse, there is a relative scarcity of terrestrial wildflowers. One can understand the plaintive query "But where were the flowers?" of Henry Walter Bates, a nineteenth-century naturalist, following a scientific trip to the Amazon. With less than one-hundredth of the total illumination at the treetops, the level of light on the rain forest floor is generally inadequate to support many terrestrial wildflowers. Nor is there any seasonal change, for there are no true seasons.

Of course, there are wildflowers in the rain forest. The great majority of showy species are hidden high in the treetops, the habitat of most of the air plants and giant woody lianas. Their aerial presence is often made known by colorful piles of fallen flowers on the forest floor. The terrestrial wildflowers, on the other hand, thrive on shade and high humidity; their foliage would be injured by wind or strong sunlight. They occur often in openings in the canopy where a giant tree has fallen or along streams that penetrate the forest. Here grow many species of *Fittonia, Pilea* and others, forming an attractive ground cover.

To utilize fully the dim light, the leaves of many of these herbs are broad and thin and often so oriented as to form beautiful natural mosaics. Such mosaics are doubly attractive when the leaves are combined with unusual shape, sheen or colorful markings. The plants that produce them, often unimpressive in bloom, are attractive for their foliage alone. Many of them, originally collected in the wild by plant explorers, have become familiar cultivated foliage plants; their natural ability to thrive in the dim light of the rain forest has made them ideal subjects for our often overheated homes. Species of *Begonia* and *Philodendron*, parlor palms (*Chamaedorea*), and marantas constitute a representative sampling of this ornamental indoor flora, members of which originally were rain forest wildflowers in some distant tropical land.

Some showy flowered species do inhabit the rain forest floor. Most characteristic is a group of large terrestrial herbs often having giant leaves, twelve to fifteen feet long, and spectacular clusters of flowers. The size of these plants seems in keeping with that of the forest trees. Most of these impressive wildflowers are mono-cotyledons, belonging to the families of the arums (Araceae), arrowroots (Marantaceae), bananas (Musaceae), cannas (Cannaceae) or gingers (Zingiber-aceae). They characterize lowland rain forests, where they form conspicuous colonies, especially in rich alluvial soil. In the paleotropics the giant herbs are most frequently wild bananas or members of the widespread ginger family, including the ginger lily (*Hedychium coronarium*), peacock plant (*Calathea makoyana*) and cardamoms. Scarlet cannas (*Canna*), arrowroot (*Maranta*) and numerous and beautiful balisiers (*Heliconia*) replace them in the neotropics. A number of these plants have utilitarian value. The cultivated bananas and plantains (*Musa*), supplying important food, originate from wild species of the Old World, while a native Philippine banana called abaca (*Musa textilis*) yields the fine fiber we know as Manila hemp.

The flowers of certain species can be bizarre.. Sometimes, as in the true cardamom of Malabar (*Elettaria cardamomum*), they appear on separate stems rising at the foot of the great leaves, while in certain Indonesian cardamoms the crimson flowers seem to be growing out of the ground itself. One of the strangest plants is the titan arum of Sumatra (*Amorphophallus titanum*); its six-foot-high inflorescence, a kind of giant tropical jack-in-the-pulpit, is one of the world's largest flower clusters. It is also one of the smelliest, relying on an overpowering stench to attract the carrion insects that pollinate it.

Colorful leaves (inflorescence bracts), rather than petals, are often the showiest part of rain forest wildflowers. Perhaps the greater surface offered by the larger colored leaves is more attractive to pollinators than are smaller petals in a dim forest habitat. The brilliant red, yellow or orange inflorescence bracts of the New World balisiers are good examples of this.

Whether erect or pendant, these bizarre inflorescences are made up of overlapping bracts. In the erect inflorescence types, rainwater accumulates within the bracts, almost covering the tiny flowers and later the cobalt-blue berries. The bright-colored inflorescences are efficient beacons for hummingbirds, which are always seen hovering around them.

The antithesis of the giant forest herbs is the group of rather insignificant little wildflowers, largely saprophytes, that inhabit forest litter. In the tropical rain forest, because decay is rapid and little humus accumulates, these plants are uncommon. They include terrestrial orchids, their near-relatives the burmannias (*Burmannia*), and some curious tropical relatives of the gentians.

43. *For an instant, a single beam of sunlight strikes pendant, beautifully formed and vividly colored florets that are bejeweled with condensation literally pouring off the surrounding foliage. The scene is the mountain cloud forest of Costa Rica, and the plant is a rare species,* Dicliptera trifurca. *It is a member of the acanthus family, whose brightly colored inflorescences are often encountered in the world's tropics.* (John Shaw)

44 overleaf. *Climbing high into the rain forests of India and Malaysia, Bengal clock vine (*Thunbergia grandiflora) *covers the leafy canopy with sheets of large blue flowers. Like many tropical wildflowers, these strong climbing lianas bloom out of sight of man, and their presence is betrayed only by piles of spent flowers fallen to the forest floor. A member of the acanthus family, it has escaped from cultivation to become naturalized in the New World tropics.* (James H. Carmichael, Jr.)

46. *Early explorers of the New World, many of them priests, saw great religious symbolism in the extraordinary flowers that spring from woody vines whose foliage forms curtains of green in South American forests. They called them passionflowers, for in their decorative fringe they saw the crown of thorns worn by the crucified Jesus, while the five petals and five sepals represented the ten faithful apostles (without Peter and Judas). Moreover, the five wounds suffered by Christ were seen in the five stamens, and the three knobby stigmas were the nails that held him to the Cross. There are some 600 species of* Passiflora, *and all have similar flowers. Because their foliage is particularly attractive to certain*

butterfly larvae, some passionflowers have developed defenses: poisonous spines or nectar glands that attract ants, which in turn kill caterpillars. (François Gohier/ Photo Researchers, Inc.)
47. *The nine-inch-wide, coconut-scented flowers of chalice vine* (Solandra maxima) *change from light yellow to rich gold as they age, hence their Spanish name* copa de oro, *or "cup of gold." The nightshade family, to which this liana belongs, provided the world with the "Irish" potato, which Spanish conquistadores brought home from the Andes in the 1500s — as well as eggplant, tomato, peppers, tobacco, the drug atropine, and deadly poisons.* (Walter H. Hodge/ Peter Arnold, Inc.)

49. *The lobster claw* (Heliconia wagnerana) *of Central America is one of a hundred or more species of* platanillo, *or "little banana," which crowd the more open floor of tropical American rain forests. The brightly colored "claws" packed on erect or pendant inflorescence stalks are actually bracts, which may be dark red, scarlet, orange, or yellow and tinged with green. The insignificant purple flowers are hidden within the bracts, often in a tiny pool of rainwater where mosquitoes breed. Hummingbirds come to heliconias for both nectar and water. These are close relatives of the bird-of-paradise flower and the edible banana, both of the Old World tropics.* (Walter H. Hodge/ Peter Arnold, Inc.)

48. *A shrub of rain-forest borders in Southeast Asia, pagoda flower* (Clerodendrum paniculatum) *grows to a height of six feet and produces a spectacular cluster of scarlet tubular flowers that appears from spring to autumn. Like a Buddhist pagoda, the layers of flowers decrease in size toward the top of the foot-high inflorescence.* (Walter H. Hodge/Peter Arnold, Inc.)

50. *The gaily colored flowers of firespike* (Odontonema strictum), *a shrub native to the wet rain forests of Central America, make it an attractive addition to tropical gardens.* (Betty Randall)

51 *top left. The red color of spiral flag* (Costus sp.), *growing in the upper Amazon Basin of Peru, suggests it is pollinated by hummingbirds. Over a period of several days the spike erupts into flower from the bottom up; this long blooming period ensures that at least some fertilization will occur should heavy rains interfere with the activity of pollinators. Like the other three plants on this page, it belongs to the aromatic ginger family, occurring throughout the world's wet tropics.* (E. S. Ross)

51 *top right. The brilliant color of red ginger* (Alpinia purpurata) *of South Pacific islands is provided by bracts; its flowers are small, white, and insignificant. The ginger used for cooking and medicinal purposes comes from an Asiatic plant no longer found in the wild.* (Walter H. Hodge/Peter Arnold, Inc.)

51 *bottom left. Hidden lily* (Curcuma australasica) *belongs to the rain forests of northern Australia. From plants of this Indomalaysian genus comes turmeric, used as a dye and a flavoring in curry sauces; East Indian arrowroot, a starch used in food preparation; cardamom, a spice used in sausages, perfumes, and incense; and zedoary, a bitter-tasting drug used as a stimulant.* (Stanley Breeden)

51 *bottom right. Torch ginger* (Nicolaia elatior) *is a giant herb native to Java and the Celebes. Like other wild gingers, its showy "flower" is really a cluster of bright-colored bracts.* (Alan Power/ Bruce Coleman, Inc.)

52 *overleaf. Truly a monster among flowers—in size, appearance, and habits—the giant rafflesia* (Rafflesia arnoldi) *of Sumatra is three feet across and weighs as much as twenty pounds. A parasite, it has no leaves, stem, no roots and thus is perfectly suited to life in the darkest rain forest. Its seed germinates in a crack in the bark of woody vines of the grape family—and grows inward! Tapping the food supplies of its host, rafflesia forms a flower bud which breaks through the bark, swells to the size of a small cabbage, and, according to some reports, opens with an explosive sound into a colossal flower whose odor of rotting meat attracts pollinating flies. Natives, who call the flower* bunga pakma, *believe a* potion made from its buds is a powerful aphrodisiac. The Indonesian government has established a nature reserve to protect giant rafflesias in western Sumatra. Twelve other, smaller species of* Rafflesia *occur in Malaysia.* (Michael Tweedie/ Natural History Photographic Agency)

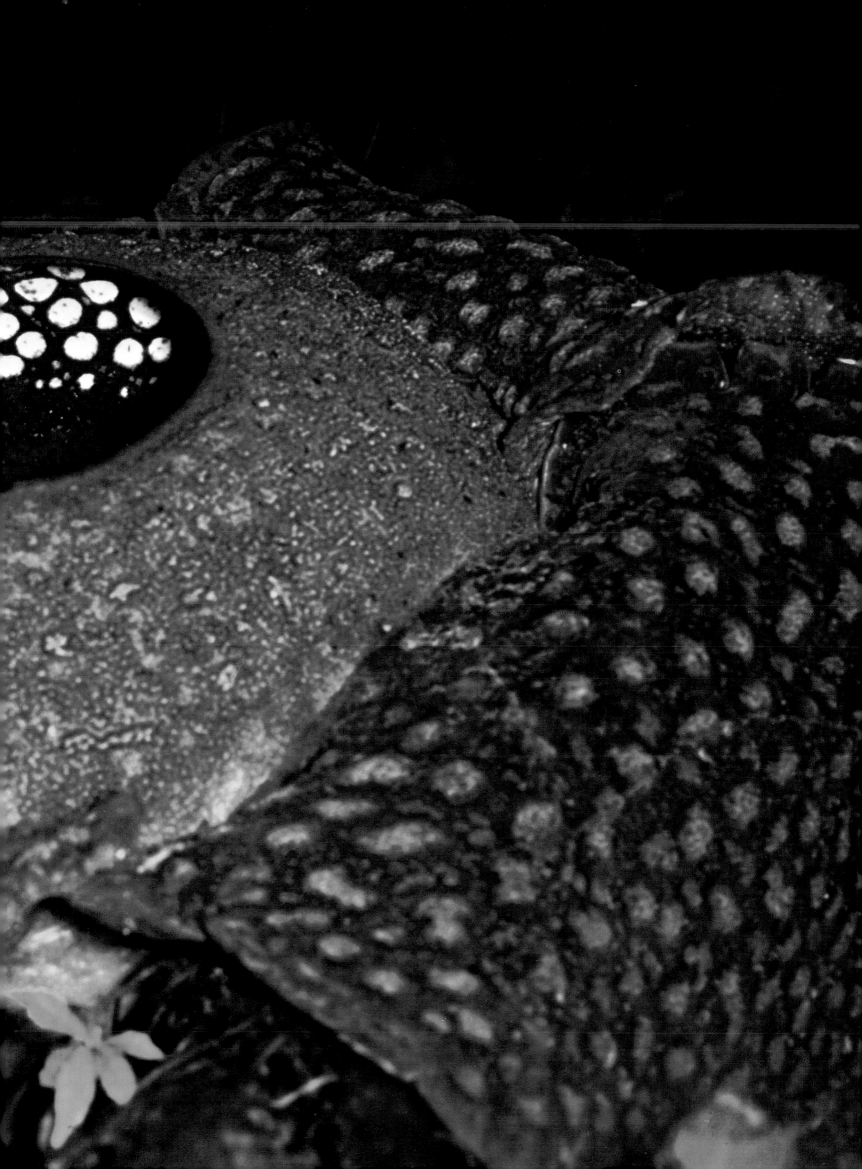

54. *Cacti of the genus* Opuntia *are found from Massachusetts to British Columbia and southward to the Strait of Magellan. They may be low-growing prickly pears or the twenty-foot tree cacti that have evolved on the Galápagos Islands; but all members of this diverse genus are easily recognized by their stems, composed of a multitude of fleshy joints. The Pueblo Indians of Arizona once cooked the cactus joints as vegetables—after removing their fearsome spines, of course. Candy and jam are still made from the tasty red fruit. A few species, such as* Opuntia rufida *of Mexican deserts, lack spines; but all have tufts of glochids—tiny barbed bristles that detach on contact, quickly and painfully penetrating the skin. This cactus is called blind prickly pear because cattle like to feed on its fruit and joints, and may be blinded if the glochids pierce their eyes.* (Farrell Grehan)

Amid Sand and Sun

I suddenly see what looks like a plucked wildflower,
a radiant yellow daisy—and still fresh! It hardly seems
likely here, amidst the barren stony plains of arid
Namaqualand. Curiously, it appears attached to a stone,
or rather it seems to be growing from a crevice formed
by two whitish pebbles. I touch the pebbles. They
prove to be turgid succulent leaves, a silver-colored
pair of them, representing a single little plant. It is
perfectly camouflaged, resembling the whitish pebbles
of the desert floor. Without its yellow daisy-sized
bloom, I would never have spotted the plant. What
I have found is a silverskin (*Argyroderma*), one of
South Africa's amazing "flowering stones." Their
mimicry enables these curious plants to survive in a
desert land where herbivores must seek every succulent
morsel that can be found.

When desert rock patches are white, the leaves of
stone plants are similarly colored, whereas reddish
soils and pebbles may be occupied by reddish-leaved
mimics. Leaves are either angular, globular or even
roughened, seemingly to match the shapes and surface
textures of the stony desert pavement. And in some
species the leaves have even evolved curious little
windows—areas devoid of green color—on their upper
surfaces which, like a skylight, permit sunlight to be
carried into the heart of the plants.

Deserts, arid lands with low annual rainfall (usually
less than ten inches), are found throughout the world
but mainly in the middle latitudes. They occur on the
continents where moisture-bearing winds, borne over
oceans, fail to penetrate; or where mountains create

rain shadows, as in the intermontane basins of western North America; or where, as in Chile and Namibia, cold coastal ocean currents withhold precipitation from adjacent lands.

In the desert, water determines which plants can live and where. If annual rains are relatively great, much of the ground will be covered with perennial plants. Where rainfall is scarcely measurable, as in the Chilean Atacama, driest of all deserts, plants are practically nonexistent. But even close to the bone-dry Atacama there exist curious hilltop oases called lomas, saturated by seasonal coastal fogs, where the desert may be temporarily cloaked with wildflowers. Depending upon their geographical location, deserts are termed either warm or cool. Cool deserts, like the Great Basin desert of North America and the similar deserts of China and the U.S.S.R., occupy the more elevated plateaus of temperate lands. These cool deserts share a similar overall gray-green look.

In America this look is produced by the scattered shrubby sagebrush (*Artemisia tridentata*), greasewood (*Sarcobatus vermiculatus*) and shadscale (*Atriplex confertifolia*); and in Asia by saltbushes, chenopods (*Chenopodium*) and wormwoods (*Artemisia*). These shrubs of cool deserts have small leaves and mostly inconspicuous flowers. Colorful wildflowers are neither as abundant nor as flamboyant as those found in warmer deserts. Moreover, they are mostly perennials. In America they include paintbrushes (*Castilleja*) and lupines (*Lupinus*), pentstemons (*Pentstemon*), balsamroot (*Balsamorhiza sagittata*) and sego lily (*Calochortus nuttallii*), in late spring and early summer; in central Asia the showy perennials are more often bulbous plants such as desert tulips (*Tulipa*), irises (*Iris*) and wild onions (*Allium*).

The deserts of warm temperate and subtropical lands are what most of us usually think of as the typical desert habitat. Temperatures in such deserts are higher than in any other habitat. Without the cooling effects of shade and abundant transpiration so typical of forests, the desert sun daily builds up abundant heat, which is rapidly dissipated at night because of the lack of an insulating blanket of atmospheric moisture. Thus deserts are blazing hot during daytime hours but often crisp and cool at night.

Inadequate and usually erratic rainfall accounts for the desert habitat. Infrequently, the annual rains may be exceptional, resulting in brilliant displays of ephemeral wildflowers. Desert ephemerals have little competition for space, so they are able to cover the ground completely between the scattered permanent vegetation. They terminate their short life cycles within a few weeks after the rains have ceased, while water in the surface soil is still adequate. Thus they escape drought completely. A few weeks after their seeds are set, they have dried up and the desert floor is again a barren waste.

To ensure continuation of the species, desert annuals must rely entirely on the viability of their seeds. These may lie dormant in the soil for years until growing conditions are just right for them. To help ensure against false starts triggered by inadequate rains, chemicals that inhibit germination have developed in the seed coats of many of these plants. Only after ample rain has fallen, sufficient to leach out these chemical deterrents, will the embryo plant begin its rapid germination.

The rainy season, when it comes, also stimulates the flowering of the perennial wildflowers of the desert. Except for the various herbaceous bulbous plants that pass through the many months of drought in underground dormancy, these perennials are taller plants whose unusual life forms produce the characteristic year-round aspect of desert vegetation. Succulents and certain trees and shrubs with small leaves, so-called microphylls, constitute the two principal perennial plant forms of warm deserts. The former, exemplified by cacti (Cactaceae) and century plants (*Agave*), are able to grow in many parts of the desert; their extensive but very shallow roots gather moisture from dew or from the lightest shower and store it in swollen leaves or stems. The microphylls, on the other hand, like the New World mesquites (*Prosopis*) and Old World tamarisks (*Tamarix*), often seek out dry riverbeds—the desert washes and wadis—where their deep-probing roots tap permanent supplies of underground water.

The development of succulent tissues for water storage is but another modification evolved by plants to survive long periods of drought. Such succulence appears either in the leaves or stems. The larger leaf succulents, like aloes and agaves, produce attractive rosettes, while stem succulents assume many bizarre forms, such as the barrels of certain cacti and the giant candelabra of African euphorbias (*Euphorbia*). Without some kind of protection, desert succulents would be great food for browsing animals. The flowering stones have resorted to camouflage, mimicking the pebbles among which they grow. Aloes and desert spurges contain bitter and sometimes poisonous compounds, making them unpalatable. But a majority of desert succulents have acquired defenses typified by a wide range of thorns, barbs and spines.

A popular misconception is that all succulents belong to the cactus family. Actually all cacti are succulents, but all succulents are not cacti. In fact, desert succulents belong to many other plant groups, including the families of the spurges (Euphorbiaceae), milkweeds (Asclepiadaceae), lilies (Liliaceae), dogbanes (Apocynaceae), agaves (Agavaceae), carpetweeds (Aizoaceae), daisies (Compositae) and orpines (Crassulaceae). Cacti are native almost exclusively to New World deserts. Desert succulents abound primarily in two parts of the world, America and South Africa. Curiously, the more conspicuous succulents of each of these regions have a very strong superficial resemblance in their habit of growth, yet they belong to unrelated families of plants. The casual traveler can easily confuse these Old World/New World look-alike succulents, especially if they are not in flower. This similarity of growth form of unrelated species is called convergent evolution.

Unlike desert annuals, desert microphylls must endure drought. Certain microphylls, such as the evergreen creosote (*Larrea*) and bur sage (*Franseria*) of Sonora, can withstand reduction of the water content of their leaves by as much as 50 percent without apparent injury. Others, such as the indigo bush and ocotillo of the American southwest, drop their leaves at the onset of the dry season, renewing them when rains begin again. On the other hand, the desert proteads and wattles of Australia long ago found an evolutionary way to combat water loss. Their leaves evolved either into slender conifer-like needles or into phyllodes, that is, bladeless leaves; both modifications offer less transpiration surface to the dry desert air. For most of the year, desert microphylls seem drab and lifeless, but after the winter rains many of them become beautiful shrubs. Curious flower clusters appear on desert proteads; the indigo bush is enveloped in blue mist, while the wattles of Australia brighten desert landscapes with mounds of gold.

59. Mariposa *is Spanish for "butterfly," and mariposa lilies are the brilliantly colored butterfly flowers of the arid American West. "They are among the most beautiful of wildflowers," one botanist commented, "and among the most astonishing, many with elaborate decorations on the petals." Forty-six species have been described, among them* Calochortus kennedyi, *whose incredibly orange flower is made even more vivid by the relentless sun over Joshua Tree National Monument, in southern California's Mojave Desert. Color, however, is not an adequate guide to identification of mariposa lilies, for many species have two or more color varieties:* C. kennedyi, *for example, is orange in California, yellow in Arizona; the flowers of* C. gunnisoni, *found from South Dakota to New Mexico, may be white, yellow—or purple! There also are many colorful local names for mariposa lilies: cat's ears (for its fuzzy petals), fairy lantern, globe lily, goldenbowl, snowdrops, Indian bells, satin bells, meadow tulip. American Indians gathered and roasted the bulbs of mariposa lilies and passed this useful knowledge on to settlers; for Mormon pioneers, the sego lily became important survival food, and they later made it the state flower of Utah. Leaves of mariposa lilies are scant and grass-like.* (Betty Randall)

60 overleaf. *Blessed with an unusual amount of winter and spring rainfall, an alluvial wash in California's Coyote Mountains has burst into color with a memorable display of desert wildflowers. There is the white of evening primrose* (Oenothera caespitosa), *its flowers with their heart-shaped petals scattered in a dense tuft of toothed leaves; the gold of desert sunflower* (Geraea canescens), *whose seeds feed a multitude of small rodents; and the purple of sand verbena* (Abronia villosa), *its fragrant flower heads sticky to the touch.* (David Muench)

62 *all. Daisies growing from stones? That is exactly the image projected by the remarkable flowering stones as protection against plant-eating animals intent on consuming every scarce green blade in the deserts of South Africa. The "stones" are a pair of succulent leaves that mimic the color, shape, and texture of the pebbles and soil in which they grow.*

Among these are Lithops lesliei *(left),* Argyroderma schlechteri *(center), and* Lithops fulleri *(right). Plants in the genus* Argyroderma *are called silverskins, for in their resting phase during long droughts, their leaves perfectly match the patches of quartz gravel where they are confined. Each year a new pair of leaves is formed, emerging from the fissure between the old pair and absorbing their moisture.*

Stone plants in the genus Lithops *have semitransparent windows on the upper surface of their leaves, which often are buried to their tops in pebbles and sand. These admit sunlight, which passes through waterfilled cells to the chlorophyll lining the interior wall. By counting old leaf scars, botanists have determined that stone plants can live more than 200 years.* Lithops *colonies are restricted in size and often cover an area no longer than a tennis court. Their seeds pop open in the rain. (Anthony Bannister,* left; *D. C. H. Plowes,* center, right*)*

63 *all. Curious members of the milkweed family are these carrion flowers from arid South Africa: Huernia zebrina (left), Stapelia asterias (center), and Huernia guttata (right). Their flowers look and smell like decaying meat in order to attract pollinating flies.*

The succulent stems of these plants suggest the cacti of American deserts, but they lack spines; their horn-shaped pods open to release seeds rigged with the familiar silky milkweed parachutes. Often called stapeliads, after their best-known genus, they are not restricted to South Africa but are found throughout the drier parts of Africa and in Mediterranean countries, the Middle East, and India.

Some 450 species in two dozen genera have been identified. One species, Stapelia gigantea, has a flower eighteen inches across. (Kjell B. Sandved, left; D. C. H. Plowes, center, right)

64 *and* **65.** *Ocotillo, coachwhip, cat's claw, slimwood, Spanish candles, and banner cactus are all frequently heard local names for a familiar desert shrub of the American Southwest,* Fouquieria splendens. *To avoid losing precious moisture during droughts, ocotillo grows new leaves after each substantial rainfall, then quickly sheds them. Its clusters of bright-red flowers are borne at the tips of thorn-covered wands that may be fifteen feet tall; like the leaves, they may appear several times a year. The dry woody stems are burned as firewood by desert dwellers and, if cut and stuck in the ground, take root and form live corrals for horses and livestock. Desert Indians used powdered ocotillo root to relieve swelling,*

dressed new leather with the waxy coating of its stems, and ate its flowers and fruit. Ocotillo *is the diminutive of the Spanish name for a pine tree.* (Stephen J. Krasemann; Farrell Grehan)

66 *and* **67.** *The brilliance, beauty, and variety of cactus flowers—enhanced by their forbidding habitat and natural armament—can hardly be suggested by seeing only four species out of hundreds. Fishhook barrel cactus* (Ferocactus wislizenii; *66) has a massive stem that may be two feet thick and ten feet tall. Because its spines flame like torches, this species was ripped out and burned in bonfires by cowboys; the Texas legislature finally ordered a fine of $50 for their destruction. Beavertail cactus* (Opuntia basilaris; *67, top) has joints shaped like a beaver's flat tail. This species forms clumps several feet across but only a few inches high. Tree cholla* (Opuntia imbricata; *67, bottom left) has a multitude of cylindrical joints five to fifteen inches long. Chollas are notorious, for their joints break loose and become painfully attached when brushed by an animal (or human). A new plant is started wherever a joint is discarded. Growing as a cylinder perhaps six inches high, Johnson's pineapple* (Neolloydia johnsonii; *67, bottom right) has flowers that vary in color from yellow-green to magenta.* (C. A. Morgan; Bob and Clara Calhoun/ Bruce Coleman, Ltd.; Erwin A. Bauer; C. A. Morgan)

68 *overleaf. Cactus look-alikes are the spiny succulent euphorbias of African deserts, which include candelabra-shaped trees growing to a height of thirty feet. Among the best-known of its kind is the crown-of-thorns* (Euphorbia milii), *native to the semiarid southwestern part of Madagascar. Like the passion-flowers of the New World, this widely cultivated plant was viewed as symbolic of the Crucifixion. Its spiny, woody branches were identified with the crown of thorns Jesus wore on the Cross; the paired flowers were the two Marys who witnessed his death, and the scarlet bracts were drops of blood.* (James H. Carmichael, Jr.)

70. *Thousands of sunflowers (*Helianthus *sp.) greet dawn in spectacular fashion on a Nebraska prairie. The New World, particularly western North America, is the true home of the sixty species of sunflowers, but over several centuries they have been carried throughout the world. Nearly 400 years ago the English botanist John Gerard wrote: "The Indian Sun, or the golden floure of Peru, is a plant of such stature and talness that in one Sommer, being sowne of a seede in April, it hath risen up to the height of fourteen foot in my garden, where one floure was in weight three pound and two ounces, and crosse overthwart the floure by measure sixteen inches broad." Cultivated sunflowers are an important crop worldwide: their seeds yield a nutritious oil and the residue is made into cakes fed to poultry and livestock; and Americans buy thousands of tons of sunflower seeds to attract and feed winter birds. One species, the misnamed Jerusalem artichoke, is farmed in Europe for its large edible roots, which can be substituted for potatoes. (Ed Cooper)*

On Plains, Pampas and Steppe

The Argentine pampas were once one of the world's
great natural grasslands; most of these have now been
converted to agriculture. But the English naturalist
W. H. Hudson, born in a pampas home in 1846, knew
those grasslands intimately and wrote of them:
". . . before the house stretched a great plain, level
to the horizon. . . . In most places the rich, dry soil
is occupied by a coarse grass, three or four feet high,
. . . and scarcely a flower relieves its uniform everlasting verdure. There are patches, sometimes large
areas, where it does not grow, and these are carpeted
by small creeping herbs of a livelier green, and are gay
in spring with flowers, chiefly of the composite and
papilionaceous kinds; and verbenas, scarlet, purple,
rose and white. On moist or marshy grounds there are
also several lilies, yellow, white and red, two or three
flags, and various other small flowers; but altogether
the flora of the pampas is the poorest in species of any
fertile district on the globe."
Grasses are one of the most familiar families of
flowering plants, for they are both abundant and easy
to recognize. They occur in all habitats from the Arctic
to the Antarctic, including the driest desert as well
as the waters of lakes and streams. But unlike any
other plant family, they also dominate one of the
world's major plant habitats, the grasslands. These are
the treeless prairies, pampas and steppes of temperate
lands, as well as the extensive park-like savannas and
campos of the tropics and subtropics. Because of their
importance as forage plants, grasses also occupy many
a meadow and pasture, which, like your lawn, are

grasslands created and maintained artificially by man rather than by climate.

Few people consider grasses as wildflowers, for their tiny complex blossoms (florets) are seldom colorful and hence are usually inconspicuous. Since wind, a constant factor in the grassland habitat, pollinates the flowers of grasses, there is no need for them to have showy structures to attract insects. Still, the mature inflorescences are often attractive, as in the silvery pampas grass (*Cortaderia selloana*) of South America or the numerous beard grasses (*Andropogon*) of North American prairies.

The world's grasslands, usually seeming limitless rolling plains, lie between the habitats of forest and desert. Grassland rainfall is an intermediate type, insufficient to produce woodlands but more than enough to preclude desert. Like desert, most grassland habitats lie toward the center of continents, where winters are cold, summers are hot and rainfall is scanty or seasonal. The water needs of shallow-rooted grasses differ from those of deep-rooted forest trees, which require moisture for most of the year. Grasses are completely dormant in fall and winter, and precipitation in those seasons is of no value to them. But in spring and summer, grasses require adequate water. Thus grasslands are found wherever there is a season of moderate rains alternating with seasonal drought. In temperate lands this coincides with the regime of cold winters followed by warm, moist summers.

All grasslands are not alike. Luxuriant communities of tall grasses, taller than a man, usually occupy those parts of steppes and prairies that border forest country where rainfall is greatest. Where rains are minimal— on the fringes of deserts—grasses are short, often in scattered clumps, sometimes failing to cover the ground completely. Between these extremes is a mixed community of tall and short grasses.

Man has exploited the temperate grasslands of the world more than any other habitat. Formerly these seas of grass were the homes of great wandering herds of grazing animals, all living among their predators in a well-maintained natural ecological balance. Most of the wild herbivores have disappeared and have been replaced by man's own flocks and herds, which graze on what remains of the nonarable short-grass plains. On the deep rich soils formerly occupied by the taller grasses, there are now only cultivated crops. Indeed, the most productive granaries of the world—America's corn belt, the wheatlands of the Ukraine and of northwest China—occupy former tall-grass prairies. Thus it is difficult to see grasslands with their wildflowers in a pristine state.

Throughout the temperate grasslands of the northern hemisphere, the wildflowers, like those of tundra, taiga and deciduous forests, are often similar. This reflects the common origins of the northern continents and of the floras that populate them. Although the individual species may differ, one still finds anemones (*Anemone*), buttercups (*Ranunculus*) and irises (*Iris*) on each continent. But there are some wildflowers limited to each area—day lilies (*Hemerocallis*), tulips (*Tulipa*) and peonies (*Paeonia*) in the steppes of Eurasia and blazing stars (*Liatris*), goldenrod (*Solidago*) and blue-eyed grass (*Sisyrinchium*) on America's prairies.

South of the equator each of the temperate grasslands has its own distinct flora. Least abundant in wildflowers are, as Hudson said, the Argentine pampas. Not so the grasslands of Australia or of the South African veld, whose wildflower riches are well known.

Grassland wildflowers are often wonderfully adapted to the wide open spaces that they call home. For the compass plant (*Silphium laciniatum*) of the prairies, the midday sun is too intense. The leaves of this species lie on their edges, with tips pointing north and south and with the flat photosynthetic surfaces facing east and west; thus the compass-like leaves receive only the less intense rays of the rising or setting sun. Most perennial wildflowers have roots that lie deeper than the grassy turf, thus minimizing root competition. Bulbs and corms permit perennial grassland wild-flowers to survive the long dry seasons, and the great abundance of bulbous species in the South African veld demonstrates the usefulness of such adaptations. Many South African wildflowers, such as the *Watsonia*, *Gladiolus* and lilies, are also nicely adapted to fires which often race across most grasslands during the dry seasons. Such bulbous plants flower in greatest pro-fusion after a fire, perhaps because conditions for ger-mination of their seeds are favorable at such times. But their various adaptations for utilizing the wind are perhaps most characteristic of the grassland wild-flowers. While the grasses are wind-pollinated, this is not the case with showy wildflowers. However, large numbers of them, especially members of the daisy family (Compositae) rely on wind to spread their seeds. Milkweeds (*Asclepias*), dandelions (*Taraxacum*) and thistles (*Cirsium*) are familiar examples. In others, such as the various kinds of tumbleweeds (*Salsola*) and windballs (*Boophane disticha*), the whole plant or infructescence eventually breaks off and is blown across the grassland, distributing seeds at every bounce.

In the tropics and subtropics is a variation of grassland called savanna. Again one finds a climate featuring alternating dry and wet seasons; but here, with the prevailing warmth, the timing of seasons is unimportant. The distinctive features of savanna are the trees and shrubs that usually dot these grasslands, giving them a park-like aspect. African savannas, with their numerous flat-topped acacia trees and great herds of game, are familiar examples of this habitat, but extensive savannas also occur in northern Australia and throughout the New World tropics, where they are known as campo (Brazil) or llano (Venezuela and Colombia).

Cuba has many attractive savanna landscapes, and there I first became acquainted with some of the flowering plants of these tropical grasslands. The savanna trees of Cuba are usually palms, in surprising variety, but mostly found only on the flat savanna-lands among tough wiry grasses. Cuban savannas look poor and are poor, for the soils are either siliceous or serpentine, making them useless to man and domestic animals. Such land, however, is congenial to many interesting herbaceous and woody wildflowers.

Perhaps the richest variety of savanna wildflowers is found in the Brazilian campo—probably the most extensive and most varied of the world's savannas. Sometimes open and park-like, the campo is often full of small trees (*campo cerrado*) forming dry scrub woodlands. Such campo habitats provide a greater variety of niches for wildflowers than do the more open grasslands. Walking among the thick-barked trees of such campos I have found strange little trunkless palms, terrestrial bromeliads, including the wild ancestor of the pineapple, showy flowered shrubs and spectacular vines. Some of the latter, such as the flame vine (*Pyrostegia ignea*), colorful *Mandevilla* and the brilliant *Bougainvillea*, are now widely cultivated as popular ornamentals in gardens of the tropics.

75. *Downward-pointing rays and a conspicuous dome of disk flowers identify sneezeweeds, common wildflowers of prairies and fields throughout North America. Purple-headed sneezeweed* (Helenium nudiflorum) *is one of the more eye-catching of these unusual, not unattractive members of the great daisy clan. Leaves of sneezeweed once were dried and powdered to make a snuff that caused violent attacks of sneezing; according to folklore, this rid the body of evil spirits that bring on disease. Of common sneezeweed, which has a yellow "button," Mrs. William Starr Dana wrote in her charming nineteenth-century book* How to Know the Wildflowers: *"During September it is abundant in Connecticut, its bright flower-heads bordering the rivers, gilding the meadows, and illuminating many of those dim woodland pools which flash upon us so constantly and enticingly as we travel through the country by rail." The genus is named for Helen of Troy.* (Ruth Allen)

76. *When wands of goldenrod* (Solidago *sp.*) *begin to wave across the grasslands of North America, it is a signal that summer's end is near. Goldenrods belong to the daisy family, and each tiny flower has its circle of yellow rays. They are massed in showy clusters that may be flat-topped, shaped like a club or a graceful plume, or branched like an elm tree. Because they are so noticeable at a time when people suffer most from allergic reactions, goldenrods get blamed for causing hay fever. In truth, it is wind-blown pollen from the inconspicuous green flowers of ragweed that causes most of the misery. Two-thirds of the world's ninety species of goldenrod are native to the New World.* (William A. Bake)

77. *In August, fuzzy purple spikes of rough blazing star* (Liatris aspera) *dominate an Iowa prairie. Close examination with a hand lens reveals that two to three dozen florets form each of the many flower heads on the four-foot-tall plant. The edible corms, or bulbs, of blazing star were dug by American Indians and stored like potatoes for winter use.* (Carl Kurtz)

80 *overleaf. A prickly shrub of tropical American savannas, Barbados pride* (Caesalpina pulcherrima) *is widely planted as a flowering fence because of its handsome blossoms, present year-round despite long dry seasons. A member of the pea family, it grows to a height of ten feet. The origin of its name is obscure, but the species may have been discovered on that West Indian island, which had the first botanical garden in the Caribbean. Closely related is the royal poinciana from Madagascar, a tropical shade tree prized for its clusters of orange-flame flowers.* (Steve Crouch)

78. *The lupines of the Pacific coast of North America come in purple, blue, yellow, and white hues, and there is one particularly stunning flower that is rose-pink and yellow.* Lupinus nanus *is a lovely blue flower with a white standard, or upper petal, native to the Coast Range of California and the foothills of the Sierra Nevada. Botanists once recognized more than 600 species of lupines in North America, but modern taxonomists have condensed this huge array into a few dozen. One species alone was formerly split fifty ways.* (Steve Crouch)

79. *A familiar, pungent odor is the first and best clue that dozens of pretty little wildflowers of North American grasslands belong to the onion family* (Allium) *and are close relatives of the onions, garlics, leeks, and chives we buy at the market. Worldwide, there are more than 300 kinds of wild onions; this delicate lily-like flower is one of thirty species found in California alone. Though wild onion flowers come in many colors and designs, all spring from an underground bulb—which, of course, is what we eat in the commercial varieties.* (Steve Crouch)

82. *The incredible crimson of royal catchfly* (Silene regia) *seems to embody the August sun that bakes an Ohio prairie. The name of this genus is derived from a Greek word meaning "saliva;" its stems and calyx tubes exude a sticky substance that supposedly traps crawling insects intent on robbing the flowers of their nectar.* (Alvin E. Staffan)

83. *Night-flying insects pollinate many of the hundred kinds of evening primroses* (Oenothera sp.) *in North America; hence their yellow, white, or pink flowers open in late afternoon and close the following morning. Their name to the contrary, they are not primroses and bear only slight resemblance to flowers of the genus* Primula. *One large species, with petals up to two inches long, is called fluttermills because it suggests a child's toy windmill.* (Stephen J. Krasemann)
84 *overleaf. The extraordinarily beautiful sunflower called tidytips* (Layia platyglossa) *grows only on the grasslands of western California. Its broad yellow ray flowers have creamy or white tips bearing two sharp notches.* (E. S. Ross)

86. *The "eye" of black-eyed Susan (*Rudbeckia hirta) *is really dark brown—a cone of florets that provide insects with nectar. The tips of its orange-yellow ray flowers reflect ultraviolet light, to which the compound eyes of honeybees are especially sensitive. There are twenty-five species of coneflowers in North America; one lacks ray flowers altogether and holds its brown cones high against the sky to attract bees. The genus honors Swedish botanist Olaf Rudbeck, a teacher of Linnaeus who founded the botanical gardens at Uppsala.* Jon Farrar)

87. *Autumn dew sparkles on the rays of smooth aster (*Aster laevis) *on the North Dakota plains. According to Greek legend, asters were created out of stardust. Colorful stars they are, some 500 species across the Americas, Eurasia, and Africa. Most spectacular is the New England aster, with upward of a hundred rich violet rays on each flower. Growing as tall as seven feet, this species was the ancestor of garden asters.* (William A. Bake)

88 overleaf. *Golden pollen gains decorate the tiny petals of blue-eyed grass (*Sisyrinchium montanum) *in an Alberta meadow. Though its leaves resemble a tuft of grass, this is a member of the iris family. The genus includes seventy-five species scattered from Canada to Patagonia. Some have yellow, white, or orange flowers. The flowers open only on sunny mornings and disappear a few hours later.* (Carroll W. Perkins)

90. *An olive grove in Greece is a mere shadow of the forests of sclerophylls, or hard-leaved trees and shrubs, found along the Mediterranean coast many centuries ago. But this terribly abused land still explodes into a spectacular profusion of wildflowers for a month or two each spring, following the winter rains.* (Farrell Grehan)

In a Mediterranean Climate

In early spring, for a month or two, the lands bordering
the Mediterranean Sea become a wonderland of wildflowers. Following the rains of winter, many of the
shrubby hillsides, formerly drab, are covered with
mounds of yellow broom (*Spartium junceum*); dark
thickets support spectacular flowers of the rock rose
(*Cistus*); aromatic lavender (*Lavendula*) purples the
open woodlands; peonies peer from wild rocky places;
diminutive irises or brilliant anemones push delicate
blossoms from stony hillsides with hyacinths, tulips
and fritillaries nearby; grasses share fields with white
daisies (*Chrysanthemum leucanthemum*), scarlet poppies
(*Papaver rhoeas*), legumes (Leguminosae) and pink
bindweeds (*Convolvulus*); chaste asphodels (*Asphodelus*)
populate derelict land; pink oleanders (*Nerium oleander*)
hug the watercourses; and places that are occasionally
damp are claimed by little ground orchids.

Even the many ancient ruins, so widely distributed
throughout the Mediterranean, are converted in spring
to natural flower gardens. One of my memorable trips
was to Italy, where I saw the elegant temple of Ceres
at Segesta ablaze with golden daisies (*Chrysanthemum
coronarium*) and discovered wild sweet alyssum
(*Lobularia maritima*) decorating crannies of Paestum's
wonderful Doric columns and entablatures. All this
seemed quite proper, for these wildflowers have been
beloved by countless generations of Mediterranean
peoples. Some—the "lilies of the field" (*Ranunculus*)
and the "lilies of the valley" (*Hyacinthus*), myrtle
(*Myrtus*), "the rose growing by the brooks" (*Oleander*)
and "the rose of Sharon" (*Narcissus*)—are cited in

Scripture; others, such as acanthus and iris, inspired motifs for architecture and heraldry; the aromatic mints (Labiatae), lavender (*Lavendula*), sage (*Salvia*), rosemary (*Rosmarinus*) and thyme (*Thymus*), were adopted by the ancients as useful herbs.

The Mediterranean climate is unusual because the rains come in winter. Warmth and moisture, normally needed by plants in active growth, fail to coincide in the Mediterranean. But Nature has produced a special kind of vegetation that can thrive in this climate. It is an open woodland of small trees and shrubs with rather unusual evergreen leaves that are hard and stiff. They are able to function during the wet cool winters but, as protection against desiccation, during the bone-dry summers they become partially dormant. Botanists call such hard, thick durable leaves "sclerophylls," from Greek *skleros* (hard) and *phyllon* (leaf). Sclerophyll forests actually are not limited to the Mediterranean area but occur in four other widely separated parts of the world: southern California, central Chile, much of the southern fringes of Australia and the part of South Africa around the Cape of Good Hope. Each of these lands has winter rains and hot dry summers, and the same sclerophyllous look, but with a very different wildflower flora.

In a way "forest" is a misnomer for most of this vegetation, for today there is no longer a continuous cover of trees or shrubs. In the Mediterranean especially, the land, abused by man and domestic animals for centuries, is in ruin, and forests have been largely eliminated. Impenetrable thickets of tall evergreen shrubs, probably representing the bushy understory of the original woodlands, are all that remain of the forests. These bushlands are known as maquis, from a Corsican word for thicket. Similar names are used for sclerophyll forests elsewhere: matorral (thicket) in Chile, chaparral (oak thicket) in California and bush in South Africa and Australia.

When they occur, trees are found where underground water is more ample, along streams, or where rainfall is greater, on hills and mountains. The trees are usually scattered, forming open woodland. North of the equator in California and in the Mediterranean, evergreen oaks (*Quercus*) and pines (*Pinus*) are dominant; in Chile it is the soap tree (*Quillaja saponaria*); in South Africa, the silver tree (*Leucadendron argenteum*); while "down under," in Australia, where the sclerophyll forests are most lush, species of eucalyptus (*Eucalyptus*) dominate. Standing under those great eucalyptus some years ago, I couldn't help noticing how different two forests of giant trees, sclerophyll and rain forest, could be. There was nothing lush about that eucalyptus forest—no vines, no epiphytes, no constant drip of foliage. The tall shafts of the great karris (*Eucalyptus diversicolor*) were spare and clear, like so many telephone poles, and possessed what appeared to be the sparsest of leaves. Indeed, the whole sky was visible, and the brilliant Australian sun fell unimpeded on the ground. Instead of lying perpendicular to the sun, the hard leaves of eucalypts hang on edge, producing so little shade that the sclerophyllous woodlands of Australia are widely known as "shadeless forests." Unlike the dim-lit floor of lowland rain forest, the Aussie forest had a bountiful understory growth of showy sclerophyllous shrubs.

Today dense thickets with a great variety of hard-leaved shrubs and bushes are more characteristic of the world's sclerophyll habitats than trees are. All too frequently the trees have been cut off, and only the shrub layer remains. In the Mediterranean, the maquis shrubs include myrtle, broom, heath, showy rockrose, lentisk, rosemary and lavender. In the New World, the chaparral of California and the matorral of Chile have similar sclerophyllous species, but since their boundaries lie next to desert, some succulent wildflowers are also shared by these two major habitats: yucca, agave and bear grass (*Nolina*) in California, and various cacti and puyas (*Puya*) in Chile.

Just as in California, the sclerophyllous woodlands of central Chile attain their best development at moderate elevations in the coastal ranges, where the matorral has the same general look as chaparral. But Chilean wildflowers are quite different, for their relationships are with the plants of the Andes or tropical South America. Although the matorral has its own special showy flowering bulbs and numerous herbaceous species, such as the painted tongue (*Salpiglossis*) and butterfly flower (*Schizanthus*), it is best known for its attractive climbers, including the bright zarcilla (*Bomarea*), the glory flower (*Eccremocarpus*), delicate twining *Tropaeolum* and colorful *Matisia*.

But it is in the Mediterranean-type climates of South Africa and Australia that sclerophyllous woodlands produce the richest displays in the world. September is the start of spring in South Africa and brings many vernal wildflowers. Among them are the many "Cape bulbs" and members of the daisy family (Compositae), a group that often dominates the local scene, for many of the low shrubs and bushes that populate the dry plains and valleys are members of this family. Many South African wildflowers are widely cultivated, including pelargoniums (*Pelargonium*), blue lobelia (*Lobelia erinus*), countless heaths, as well as the numerous ice plants (*Dorotheanthus, Lampranthus*) and Hottentot's fig (*Carpokrotus edulis*), which here carpet local sands and flats. These are the same species of ice plants that decorate the distant roadsides of California and the Riviera with such color, but in South Africa they are native, not naturalized.

Flowering continues throughout summer and fall, though not at the dizzy pace of spring. Summer is the principal season for the innumerable heaths (*Erica*) and the succulents, while in May, the equivalent of the northern hemisphere's October, many species of surings (*Oxalis*) color slopes with bright yellow, pink or white blossoms. Fall and winter are the main flowering seasons of the shrubby proteads (*Protea, Mimetes, Leucospermum*), whose bizarre blossoms are the most unusual of South African wildflowers.

I spent a spring seeking West Australian wildflowers of possible interest as ornamentals for the gardens of America. For most of the year the rolling bushland is barren and uninviting, and the landscape is relieved only by eucalypts and the curious grass trees or blackboys (*Xanthorrhoea*). Spring is another story, with its countless wildflower treasures, totaling over 7,000 different kinds! Some of the most promising candidates for introduction proved to be shrubby members of the *Protea* family (Proteaceae), fragrant *Boronia*, *Verticordia* of various colors and the blue hibiscus (*Alyogyne huegelii*). Numerous ground orchids were found among the bushes, while on patches of sand were appealing sundews (*Drosera* and *Byblis*), trigger flowers (*Stylidium*) and kangaroo paws (*Anigozanthos*). The most showy floral displays develop in larger openings in the bushlands. There was that purest of flower colors, the blue of *Leschenaultia;* and where cleared land lay fallow, there were acres of pink, white or yellow everlastings (*Helichrysum*).

95. *When rains come to the deserts of central Australia, masses of billabuttons (Leptorhynchus sp.) burst into flower from seeds that have lain dormant in the barren sands through long periods of drought. Such plants, called ephemerals, are able to grow and bloom within a brief time. The popular name has the same native origin as billabong, a stream that usually is dry but runs during the rainy season. Rainfall in the Australian desert is erratic at best; the region may be watered by either of two weather systems, in winter or summer, and some areas go for years without significant precipitation. When rain does arrive, it may fall in barely measurable amounts. This may encourage plants to start growing, but they perish before producing seed if no additional rains follow immediately. Eucalypts are a typical tree of Australian deserts, for their vertical leaves avoid the sun's direct rays and thus minimize loss of scarce water. (Stanley Breeden)*

96 and **97**. *The spectacular inflorescence cylinders of the Australian trees and shrubs of the genus Banksia may be eighteen inches tall and contain thousands of densely packed flowers arranged in a spiral that blooms from the bottom up. This genus includes fifty-one species, some of which reach New Guinea; their flowers occur in all shades of red and yellow, from deep crimson to a green-tinged white. Banksias are pollinated by Australia's endemic honey eaters; when a bird lands on the flower spike to drink nectar, its feathers, feet, and beak are dusted with pollen that is then carried to the next inflorescence. Small marsupial mammals also frequent banksias for their nectar and the insects these flowers attract.*

The flowers represented here are Banksia ericifolia *(96) and* Banksia attenuata *(97).* (Stanley Breeden; Eric Chricton/Bruce Coleman, Ltd.) **98** *overleaf. Official flower of the Australian state of New South Wales, the waratah* (Telopea speciosissima) *has a five-inch flower head of brilliant crimson. Waratah— a native name—belongs to a southern hemisphere family of trees and shrubs called Proteaceae; more than half of its 1,200 species are found in Australia. Because their flower clusters exhibit so many different forms, the family was named for Proteus, in Greek legend a prophet famous for his power to assume any shape at will.* (Walter H. Hodge/Peter Arnold, Inc.)

100 and **101.** *Various genera of the amazing Proteaceae family are represented in this collection of colorful flowers. From Table Mountain overlooking Cape Town, South Africa, come two species belonging to the genus* Protea *(100 left, 100 center); the flower heads of the eighty species in this genus are as large as eight inches across —and almost as high. Some are called "sugarbush" because nectar can literally be poured from the huge flower heads and boiled like the sap of North American maple trees to make a sweet syrup.*

Dryandra proteoides *(100 right)* and Dryandra speciosa *(101 center) are two of forty-eight species of golden- or bronze-flowered shrubs restricted to southwestern Australia. With their stiff, wire-like floral structures and burnished surface, one Australian naturalist wrote, the dryandras give "the impression that each is an exquisite sculpture in metal rather than a living flower."*

Another native of the island continent, Grevillea refracta *(101 right) represents a genus famous for its long clusters of bright-colored flowers. Among its 170 relatives are tall timber and shade trees that are welcome in a hot land where the abundant eucalypts cast no shadows. Part of the chaparral-like flora of coastal South Africa, but a member of the composite or daisy family, is* Helichrysum sesamoides *(101 left).*

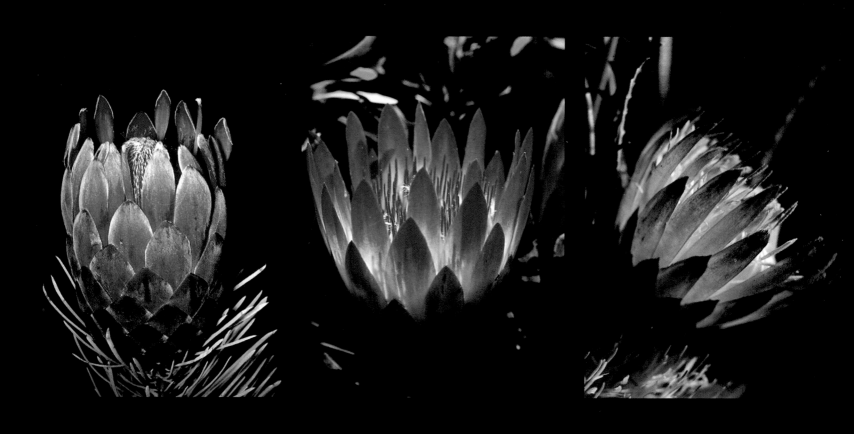

Plants in this widespread Old World genus produce the tiny strawflowers, also known as paper daisies or everlasting, sold dried and dyed in florist shops. *(100 left and 101 left,* E. S. Ross; *100 center,* Walter H. Hodge/Peter Arnold, Inc.; *100 right, 101 center and right,* Michael Morcombe)

102 *overleaf. Grown in gardens around the world, the Barberton daisy* (Gerbera jamesonii) *is found wild only in the South African province of the Transvaal. Colors of the rays of the wild species vary from this incredible scarlet to shades of orange and pinkish-yellow, and horticulturists have produced an even wider range of hybrids. This genus counts eighty species in Africa, Madagascar, India, China, and South America. The Barberton daisy, which seeks out shaded places on rocky hillsides, is named for a mining town in the Transvaal goldfields. Flowers of the daisy family often dominate the spring-time landscape in South Africa.* (James H. Carmichael, Jr.)

104 *second overleaf. Member of the parsley family and kin of wild carrot,* Tordylium apulum *is eaten as a vegetable in Greece. Growing in fields and wasteland throughout the Mediterranean area, it bears five to eight umbels of flowers. The strange outer flowers have only one large petal with two deeply cut lobes.* (Farrell Grehan)

106. *Dawn in the Okefenokee Swamp of Georgia finds the flowers of the fragrant water lily* (Nymphaea odorata) *opening against a backdrop of cypress trees hung with Spanish moss. The life of a single flower is three to four days; once it is pollinated, the stalk is withdrawn and the seed matures at the bottom of the pond. Local people call this plant "alligator bonnet," for these great reptiles often wait motionless with their bodies submerged and their heads camouflaged by aquatic plants.* (Patricia Caulfield)

Floating Gardens

During the early 1900's one of the world's most sought-after wildflowers was the giant water lily (*Victoria amazonica*) of tropical South America. Early travelers had reported the existence of this remarkable aquatic, and so for nearly a quarter century plant hunters vied to see who could first bring this wonderful flower into cultivation. One of these travelers, Richard Schomburgk, wrote in 1847 of one of his Guiana encounters with this striking plant: "It was the magnificent Victoria . . . with its rounded leaves from five to six feet in diameter and its beautiful huge flowers: the petals . . . merged from white through a series of the softest of tints to a moist rosy red, and filled the whole neighborhood with their lovely scent. . . . I bent down . . . to break off so wonderful a blossom when, as if bitten by a tarantula, we speedily withdrew our hands . . . receiving a fairly painful lesson from the three-quarter-inch-long sharp spines. . . ."

Years later, after they were established in cultivation; much more was learned about the two species of giant Victoria. The flowers, up to eighteen inches across and with a rich pineapple-like fragrance, are nocturnal, opening in late afternoon and closing the following mid-morning, and changing in color on successive days from creamy white to pink or crimson. The giant platter-like leaves possess an amazing girder-like framework of veins of such strength that a man can be supported by a single leaf. The remarkable vein structure of the great leaves even inspired the design of London's nineteenth-century Crystal Palace, the largest conservatory ever built.

Water lilies, like the giant Victoria, are but one of many kinds of aquatic wildflowers that form the familiar belts of vegetation ringing the margins of most lakes and ponds. Water lilies (*Nymphaea*) occupy the middle belt. Beyond them is an outermost belt of submerged wildflowers where hornworts (*Ceratophyllum*), pond-weeds (*Potamogeton*), bladderworts (*Utricularia*) and similar species pioneer in the deeper water. On the landward side of the water lilies, in the shallowest water near shore, lies a zone of emergent aquatics whose leaves and flowers are held well above the water.

The submerged plants, though at times rooted, are as often free-floating. The depth at which they can live is based on the clarity of the water, for like all green plants they require sunlight in order to thrive. Although they can also live in shallower water, the shading from the leaves of the surface-floating aquatics of the middle zone makes it difficult for them to compete. Similarly, water-lily-like plants find it not so easy to grow close to shore because of the shade and root competition of the shallow-water species.

Being completely covered with water has advantages and disadvantages. Submerged plants have little need for water-gathering roots or the specialized conducting systems required by terrestrial plants to transport such water, for all their parts can absorb water. Nor do these plants require special tissues to support them, for they are buoyed up by the water. But special problems exist, such as how to obtain the gases required for respiration and photosynthesis. Such gases, though dissolved in water, are in short supply. To offset this shortage, underwater plants have developed much greater plant surface areas to facilitate the absorption of gases. This explains the finely dissected, lattice-like or ribbon-shaped leaves found in many submerged wildflowers. Just as critical in an underwater habitat is the problem of flowering, including cross-pollination. Many submerged plants rely on flying insects. Water buttercups and bladderworts, for example, produce aerial stalks to raise their attractive little flowers above the water surface. Others, particularly members of the frog's-bit family (Hydrocharitaceae), have flowers especially designed for water pollination.

The best-known example of this is freshwater eelgrass (*Vallisneria*). Like numerous other flowering species, eelgrass plants are either male or female. The female plant sends up its flowers singly on long string-like cords to the water surface. There the little blossoms open, each thrusting out a trio of tiny stigmas that are held parallel to the water. At the same time the male plants produce tiny pollen-bearing flowers. These break off underwater, rising to the surface as individuals, and there open. Like tiny boats, the flowers then float until, with luck, they lodge against the female blossom. Pollination then takes place. Once pollen is transferred, the cord-like stalks of the female flowers gradually retract like springs, pulling them to the pond bottom, where seed is eventually formed.

Members of the water lily family, which include the showiest of aquatic wildflowers, have worldwide distribution. Wherever they occur, they dominate the belt of floating vegetation that characterizes the margins of ponds and quiet backwaters of streams. Lying as they do directly upon the water, the leaves of water lilies have evolved special features: The tiny breathing pores found on the lower surfaces of most plant leaves are here limited to the upper surfaces, thus permitting ready contact with the air. As with most aquatics, moreover, the upper leaf surfaces are highly waxed, permitting water to roll off easily so as not to clog their pores. Flowers of the true water lilies range in size from the tiny blossoms of the pygmy water lily (*Nymphaea tetragona*) to the plate-sized blossoms of the giant Australian species (*Nymphaea gigantea*). Wild water lilies have a variety of shades, from white and yellow to pink, red, blue and purple. Utilizing this natural color range, plant breeders have produced numerous hybrids, the water lilies most commonly cultivated.

The water lilies of the temperate region are usually day bloomers, and their flowers normally rest on the water surface; in tropical water lilies the flowers rise above the water, and one finds both day- and night-blooming species. To better attract their nocturnal pollinators, night bloomers are usually white as well as fragrant. But most sweet-scented of all is the common white water lily of eastern North America (*Nymphaea odorata*).

The water lily is the national emblem of Egypt. There the two common wild species (called Egyptian lotus, but not to be confused with the true lotus, *Nelumbo*) have been revered for over 5,000 years. Originally associated symbolically with the river Nile, the provider of life, water lily blossoms were votive offerings and appeared frequently in ancient paintings. The characteristic forms of the bud and flower served also as models for the Egyptian architectural lotus capital, a design adapted by artists of later civilizations.

Resembling water lilies are the cow lilies or spatter-docks (*Nuphar*), which often grow near their more attractive relatives in temperate areas. Spatterdock flowers are yellow or reddish and, though conspicuous, are not showy. Best known of water lily allies are the species of lotus, the yellow lotus or water chinquapin (*Nelumbo lutea*) of North America, and the sacred pink lotus of Asia (*Nelumbo nucifera*). Lotuses differ from water lilies in having both leaves and flowers held above the water.

The Orient, that is, from India to Japan, is lotus land. Throughout that region the lotus flower, signifying purity, is held sacred by Buddhism. The religious symbolism suggests that just as the lotus, among the loveliest of wildflowers, can rise from a lowly habitat, namely the mud, even so can man surmount all adversity. Buddha is frequently portrayed seated on a lotus blossom. To sit similarly on a lotus throne is the goal of his faithful followers, for that act signifies rebirth in paradise. Thus lotus flowers are standard offerings at many Buddhist temples. Because the open flowers are too delicate to transport, each bud is opened by hand to yield a fresh lotus-blossom throne for the devout. Lotus flowers open at dawn, and so a favorite Japanese summer pastime is to visit a pond at sunrise for lotus viewing.

Frequently associated with lotuses and water lilies are tiny floating plants, the duckweeds (*Lemna*). They are the most diminutive of all wildflowers. Some kinds, such as watermeal (*Wolffia*), are so small that the individual plant is scarcely visible to the naked eye. Duckweeds reproduce so rapidly and so efficiently by vegetative means that seed is seldom needed, and hence flowers are only rarely produced. Because of their size, duckweeds become easily attached to waterfowl, through whose migrations these curious floating aquatics have been widely dispersed to all parts of the world.

111. *For Buddhists from India to Japan, the sacred lotus* (Nelumbo nucifera) *has great religious symbolism. If such a lovely flower can rise from the mud, it is reasoned, then man can conquer any adversity. For botanists, however, it is the lotus seed, not its flower, that holds great interest—for it can retain life for centuries. Lotus seeds that had lain dormant in a Manchurian bog for 1,000 years germinated when they were notched to admit water and planted. There are but two species in this genus; the water chinquapin* (Nelumbo lutea) *of North America has a yellow flower eight inches across, making it the largest flower on the continent. (Walter H. Hodge/Peter Arnold, Inc.)*

112 *overleaf. Found in ponds and quiet streams throughout the world, water-shield* (Brasenia schreberi) *is named for its floating, shieldlike leaves. Submerged parts of the plant are coated with a transparent jelly that makes them slippery to the touch. American Indians relished its starchy roots, and in Japan the young leaves are eaten in salads. (Alvin E. Staffan)*

114 *second overleaf. Of the giant water lilies that inhabit the back-waters of tropical and subtropical South America, a nineteenth-century botanist-explorer wrote: "The impression the plant gave me, when viewed from above, was that of a number of green tea-trays floating, with here and there a bouquet protruding between them." The Santa Cruz water lily* (Victoria cruziana) *is found in northern Argentina, Paraguay, and Bolivia; its night-blooming flowers are pure white when they first emerge, changing to pinkish-red in succeeding days. Equally large is a species that grows in the Amazon and Orinoco basins; its leaves may reach six to nine feet across and can support a small child. Notches in the leaves allow rainwater to drain out. (Walter H. Hodge/Peter Arnold, Inc.)*

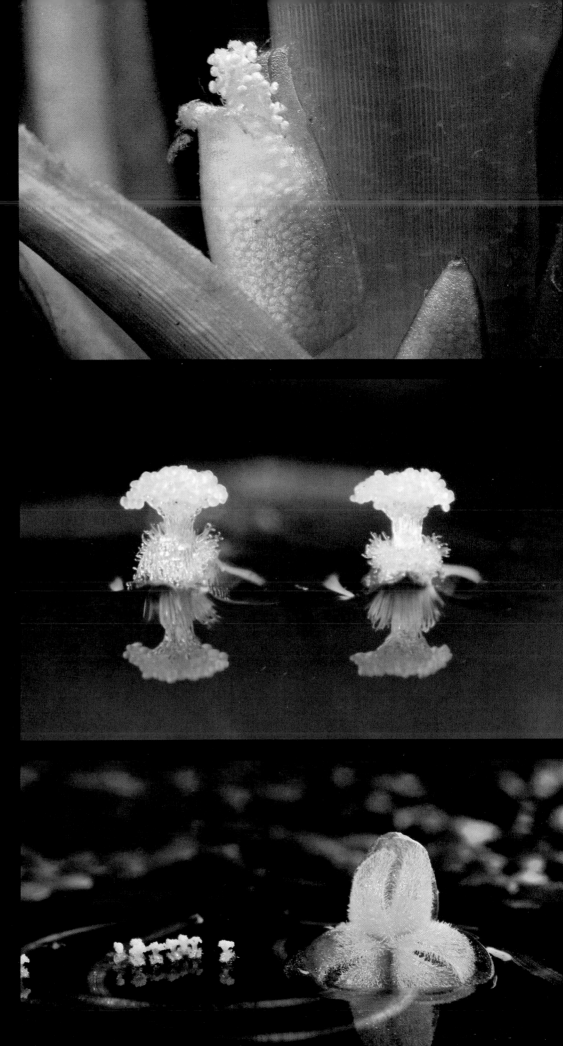

116. *Buckbean or bogbean (*Menyanthes trifoliata*) is a water-loving member of the gentian family that grows in shallow, boggy ponds throughout the cooler regions of the northern hemisphere. The leaves, rising from a horizontal underground stem, sometimes form dense floating platforms. The rootstocks once were ground up by Finns and Laplanders to make a nutritious, if bitter-tasting, bread.* (Ingmar Holmåsen)

117. *A remarkable mechanism for aquatic pollination has been devised by freshwater eelgrass (*Vallisneria spiralis*). Female flowers rise to the surface of a pond on slender stems that may be ten feet long. The tiny male pollen-bearing flowers, meanwhile, break loose underwater, float to the surface, and are gathered by surface tension about the female flower. When contact and pollination occur, the stalk of the female flower recoils and is pulled back to the pond bottom, where the seed is formed.* (Oxford Scientific Films)

118 *overleaf. Yellow floating heart (*Nymphoides peltata*) suggests a small water lily with heart-shaped leaves, and in its native Europe it is erroneously called fringed water lily. But, like buckbean, this is an aquatic member of the gentian tribe.* (M. P. L. Fogden)

120 *second overleaf. The buttery flowers of the aquatic plant known variously as spatterdock, cow lily, and bullhead lily (*Nuphar luteum*) decorate still waters across Canada and the northern United States from May to October. The open globe of the flower, which has an odor of alcohol, is formed by the leathery sepals. Roots of this plant were an important food to American Indians, who often robbed muskrat houses of the rodents' stores.* (Susan Rayfield)

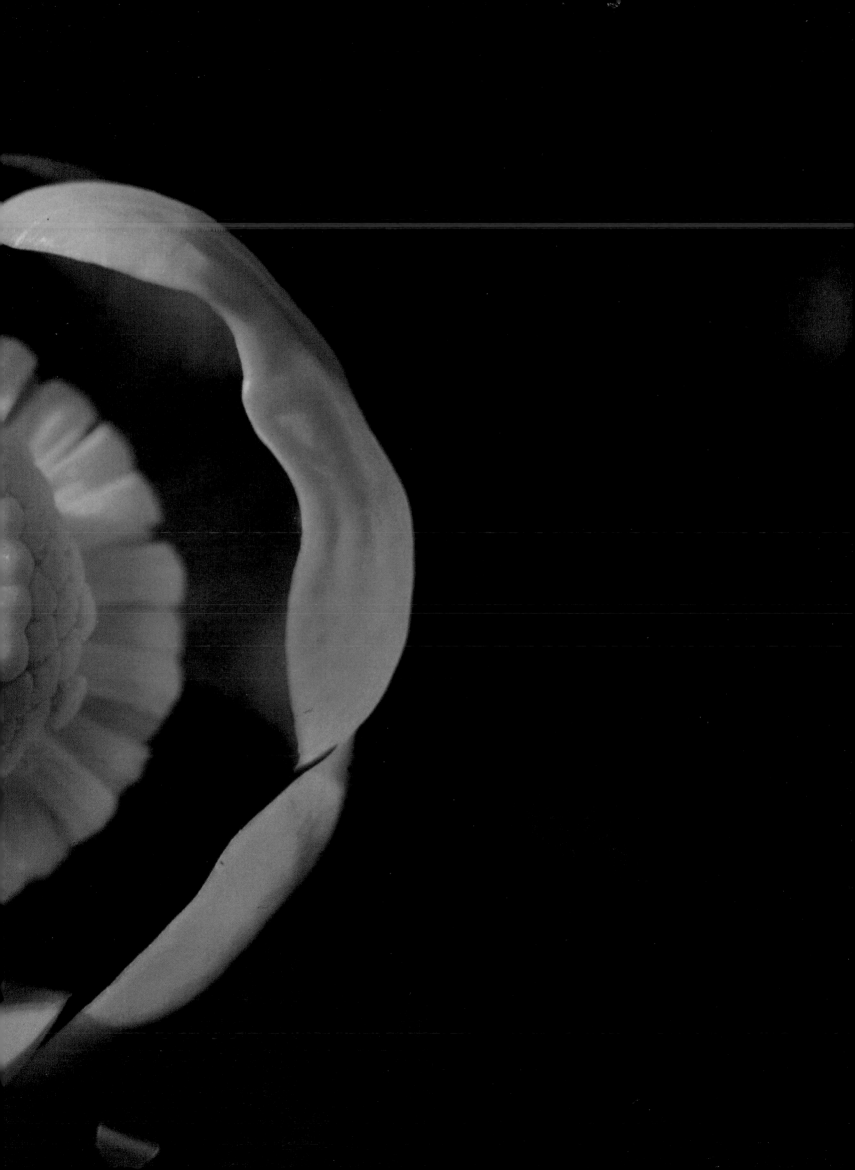

122. *A spring marsh in Japan is beautified by the white spathes of an East Asiatic skunk cabbage* (Lysichiton camtschatcense) *and the butter-yellow flowers of marsh marigold* (Caltha palustris). *The latter is not a marigold, however, but a water-loving buttercup that is a familiar early flower in swampy places across North America and Eurasia. Its goblet-shaped leaves are a delicious potherb. The spectacular part of the skunk cabbage display is its spathe, which appears before the leaves develop; the minute flowers are grouped on a thick stem, called a spadix. Skunk cabbage belongs to the arum family, a large and mainly tropical group of plants that includes the familiar philodendrons found in millions of homes and also taro, which feeds millions of people in Southeast Asia and on Pacific islands. One incredible plant, the titan arum* (Amorphophallus titanum) *of Sumatran rain forests, bears a spathe that can be 8½ feet tall—one of the largest floral clusters in the world. Legend has it that elephants drink rainwater from the plant's four-foot-wide spathes and carry pollen from plant to plant on their brows, like giant leathery bees, but insects are actually the pollinating agents.* (George Holton)

Between Land and Water

Swamps and marshes, both large and small, are familiar wetland habitats. They exist wherever drainage is poor and where standing or running water is commonplace. The Florida Everglades, a vast area of hundreds of square miles of freshwater marsh, wooded swamps and coastal mangrove thickets, unite some of the major variations that one finds in the wetlands of the world.

The Everglades cover land that is flat and practically at sea level. Thus what remains of the natural overflow that pours southward over this essentially treeless land from Lake Okeechobee is slow and sluggish. The traveler crossing the Everglades is everywhere conscious of the poor drainage. Roads are flanked with water-filled ditches often clogged with the blue of water hyacinths (*Eichhornia crassipes*) or with masses of pioneering cattails (*Typha*). In all directions are waving seas of tall saw grass (actually a sedge, *Cladium jamaicensis*), which forms a turf so tight that few other plants can compete with it wherever it is established. Here and there are open channels through which flows the slow-moving water. Occasionally dotting these great wet saw-grass marshes are small raised islands of trees, shrubs and palmettos called hammocks.

Like most marshes, the Everglades are rich in wetland wildflowers, although one seldom sees great masses of color. To be sure, the prairie flag (*Iris hexagona* var. *savannarum*) sometimes forms huge colonies on the borders of these marshlands, but more typical are the occasional clumps of white arrowhead (*Sagittaria*), golden canna (*Canna flaccida*), or blue pickerelweed (*Pontederia cordata*); the rose of marsh pink (*Sabatia*) or swamp hibiscus (*Hibiscus moscheutos*); and patches of fragrant swamp lily (*Crinum americanum*).

Sunny grassy freshwater marshes not unlike the Everglades—except for the kinds of wildflowers found in them—occur in many parts of the world. They occur most frequently at the mouths of rivers, as in "Las Marismas" of Spain's Guadalquivir, the Rhone's Camargue, the Mississippi bayous, and similar delta marshes on all continents. Aggressive grasses, sedges, cattails or rushes usually dominate such freshwater marshlands, often as great masses of a single species. Although it is the common Caribbean saw grass in the Everglades, elsewhere it may be another species. Thus the cosmopolitan reed grass (*Phragmites*) covers the huge Danube delta, while papyrus sedge (*Cyperus papyrus*) jams the vast morass of the Sudd on the Upper Nile; but two-mile-high Lake Titicaca in Peru, on the other hand, is fringed with colonies of bulrushes and cattails.

Unlike marshes, swamps are usually wooded habitats, and are more or less shaded rather than open and sunny. The kinds of trees that can thrive successfully with roots periodically inundated are few, and so the species of trees dominating swamp woodlands are relatively few. West of Florida's marshy Everglades is Big Cypress Swamp, which characterizes such habitats. This is not high ground but rather lowland swamp, full of such typical trees as water oak (*Quercus nigra*), tupelo (*Nyssa sylvatica*) and especially bald cypress (*Taxodium distichum*). The first trees we meet are often smallish. Many have epiphytic bromeliads perched on the sides of their slender trunks. Once well within the swamp, the water deepens and the trees increase both in height and girth, becoming now mostly cypress with flaring trunks and branches bedecked with gray Spanish moss. Although this is a woodland swamp, there are numerous open watery glades, and in these one finds the wildflowers of the swamp. Many, such as water lettuce (*Pistia stratiotes*), water hyacinths, pickerelweed and arrow arum (*Peltandra virginica*), are shared with the open Everglades.

The cypress-tupelo swamp is widespread in the lowlands of the North American coastal plain from Delaware to Texas. In the tropics are similar swamps but with quite different trees and wildflowers. The latter are often giant arums (*Araceae*) called elephant ears, traveler's palms (*Ravenala madagascariensis*), and other large monocotyledonous herbs. The *igapó*, the continuously inundated woodlands that fringe the great rivers of Amazonia, is just such a swamp forest of the lowland tropics. In cool north temperate regions, swamps are often dominated by deciduous trees. There the showiest swamp flowers are often spring-flowering. Depending on whether the swamp is in Europe, Japan, or America, one may find buttercups (*Ranunculus*) and marsh marigolds (*Caltha palustris*), primroses (*Primula*), ragworts (*Senecio*), violets (*Viola*), plantain lilies (*Hosta*) and false hellebores (*Veratrum*), and species of the skunk cabbage (*Symplocarpus*).

West of Big Cypress Swamp lies the Gulf of Mexico. Along its shores south to the Florida Keys is another kind of tree-dominated swamp, that of the curious mangroves (*Rhizophoraceae*). This is a saltwater rather than a freshwater habitat. In one species or another, stilt-rooted mangroves form impenetrable land-claiming woodlands that fringe the saline tidal flats of most coasts lying within the tropics. In their ability to grow and pioneer colonies in shallow seawater, mangroves are rarities among trees. In fact, few flowering plants of any kind can exist under such difficult conditions. Because of this, herbaceous wildflowers, except for occasional epiphytes or submerged colonies of marine grasses, are seldom seen in mangrove swamps.

No mangrove-like trees have evolved in the more rigorous climates of temperate zones. There the closest counterpart of the mangrove swamp is the familiar salt marsh. The salt marsh develops, in sheltered estuaries along seacoasts, wherever streams meet the sea. Turf-forming cordgrasses (*Spartina*) usually dominate it, thriving in waters that change daily with the tides from salty to brackish. Though seemingly useless, salt marshes and mangrove swamps are ecologically among the world's most important and productive plant habitats. Organic materials produced in these extensive tracts are daily washed into the seas, serving as the basis for the food chains on which the world's fisheries depend.

Bogs are special kinds of peaty wetlands that differ greatly from marshes and swamps. They occur in cool temperate, boreal and subarctic lands, primarily within the realm of conifer forests. There the land is dotted with glacial ponds and lakes. Many of these, especially the so-called bog ponds, have been invaded by peat mosses, which over the years ultimately fill the pond basin, permitting eventual restoration of conifer forest.

Visiting a bog is a unique experience for the wildflower lover. My favorite time for such a visit is in late June or early July. We approach the hidden bog through a fringe of tall conifers—spruce and balsam. Occupying what years ago was the original pond margin, these trees are now tall, fully mature; but as we make our way beneath them, they gradually decrease in size until they are no taller than a man. At this point the open water, the "eye" of the bog, is usually visible. It may be no more than a hundred feet across, depending upon how far the successional or filling-in process has progressed. Here the bog becomes really interesting, in more ways than one. I usually make a big jump or two; this causes all the vegetation round me to tilt and sway, for we are now on the quaking part of the bog. This great floating mat of spongy sphagnum moss is interlaced by countless roots of associated shrubs and herbs. One uses caution in approaching the visible shore, sometimes rather tipsily, for at its edges the mat underneath us is thinner, and there is always the possibility of falling right through.

The waters of a northern bog pond are brown, acidic and low in mineral content. On the open water there are often white water lilies (*Nymphaea*) or yellow spatterdocks (*Nuphar*), with perhaps buckbean plants (*Menyanthes*) closer in. But the wildflowers found on the bouncy floating mat of sphagnum are what we have really come out on the bog to see. Most only grow on such bogs and hence are often rare. A site like this is the home of some of the most beautiful or unusual plants: choice orchids such as the showy yellow lady's-slippers (*Cypripedium pubescens*), various fringed orchids (*Habenaria*), as well as the lovely *Pogonia*, *Calopogon* and *Arethusa*; curious carnivorous sundews (*Drosera*) and the northern pitcher plant (*Sarracenia purpurea*). Of equal interest are shrubby heaths—cranberries (*Vaccinium*), leatherleaf (*Chamaedaphne calyculata*), bog laurel (*Kalmia polifolia*), Andromeda and Labrador tea (*Ledum groenlandicum*). Therefore, part of the thrill of seeing these bog plants is realizing that some of these wildflowers cannot be seen elsewhere, except in lands far to the north.

127. *One of only two species in its genus, western skunk cabbage (*Lysichiton americanum*) has a bright-yellow spathe that decorates spring woods from northern California to Alaska and eastward to the Rocky Mountains. There is a purpose to the foul odor of skunk cabbage, which becomes particularly obnoxious if the leaves are crushed: it lures the carrion flies that cross-pollinate the flowers. American Indians dug and roasted skunk cabbage roots and ground them into flour. Bears also are fond of this and other arum roots found across northern Europe and Asia. Hence came a fable reported as fact by such early scholars as Aristotle and Pliny, and repeated as late as 1597 by the English botanist-surgeon John Gerard, who wrote: "Bears, after they have lain in their dens four days without any manner of sustenance but what they got with licking and sucking their own feet do, as soon as they come forth, eat the herb Cuckoopint; through the windy nature thereof the hungry gut is opened and made fit again to receive sustenance, for by abstaining from food so long a time the gut is shrunk or drawn so close together that in a manner it is quite shut up." (Walter H. Hodge/ Peter Arnold, Inc.)*

128 *overleaf. The yellow flowers of cinquefoils are known throughout northern lands; there are some 500 species in the genus Potentilla, of the rose family. The common name, derived from the French, means "five-leaved"; the generic name translates as "little potent one," because of the reputed medicinal powers of a tea brewed from the foliage. Shrubby cinquefoil (*Potentilla fruticosa*) is a woody, bushy species found in bogs and meadows across Eurasia and North America. (C. W. Perkins)*

130 *second overleaf. The stringy flowers of swamp lily (*Crinum sp.*) glow in a mangrove swamp on the Mexican coast. Equipped with a corky covering, Crinum seeds have floated across oceans to colonize tropical and subtropical shores throughout the world. Botanists recognize some one hundred species in this genus, which belongs to the amaryllis family. (M. P. L. Fogden)*

132 *and* **133.** *The fringed orchids
are among the most beautiful
flowers in the genus* Habenaria,
*which numbers some 600 species
worldwide. The individual flowers,
clustered on a stem that may be
four feet high, have a fringed lip
a half-inch long. From the base of
the lip a hollow spur, containing
a drop of nectar at its tip, extends
backward. Petals and a sepal form
the hood. In a southern Michigan
sphagnum bog, a white fringed
orchid* (Habenaria blephariglottis),
*above, grows side-by-side with a
hybrid, opposite page, between this
species and the yellow fringed
orchid. In some bogs, hybrids out-
number the parent species, and
every kind of color combination
occurs.* (Jack Dermid; Larry West)

134 *and* **135.** *Among the more
brilliantly colored of the native
North American terrestrial orchids
is the yellow fringed orchid*
(Habenaria ciliaris). *Its rich orange
flowers are found in a variety of
habitats—bogs, damp meadows,
pine barrens, even upland woods—
from Massachusetts to Wisconsin
and southward to Florida and Texas.
Fringed orchids have an unusual
method of ensuring a plant's year-
to-year survival. Each season the
tuberous root produces a bud that
develops a new set of roots. When
the old plant degenerates at the end
of the flowering season, its replace-
ment is waiting in the wings.*
(Larry West)

136 *and* **137.** *Another favorite of orchid hunters, the purple fringed orchid* (Habenaria psycodes) *can be discovered from June to August on swampy lakeshores, along creeks, even in roadside ditches and wet pastures. On some northern prairies it grows so abundantly that the grassy expanses take on a purple hue. Fringed orchids are pollinated by butterflies and moths. Hovering in front of a flower, the insect will unroll its long tongue and thrust it into the spur to reach the nectar at its bottom. In the process the orchid's pollinium— a tiny wad of pollen—becomes stuck to the insect's tongue and is carried to a second flower.* (Susan Rayfield; John Shaw)

139. *A skipper butterfly has spent the night clinging to a grass-pink* (Calopogon pulchellus) *in a Michigan bog, and at daybreak both insect and flower are dew-covered. The generic name of this lovely little orchid is derived from two Greek words meaning "beautiful beard." Unlike typical orchids, the large petal or lip of this species stands erect at the top of the blossom. To a bee gathering pollen, the fringe of yellow hairs on the lip—the beard—suggests a mass of stamens. When the bee lands, the petal bends and the insect slides onto the pollinia, which stick to its back. Cross-pollination occurs when the bee repeats this quite comic routine on another flower.* (John Shaw)

138. *Of the exquisite arethusa* (Arethusa bulbosa), *whose face-like flower also is known as dragon's-mouth, two orchid specialists wrote a half century ago: "We shall never forget the moment when our eyes first fell on its blossom in the lonely depths of a sphagnum bog. The feeling was irresistible that we had surprised some strange sentient creature in its secret bower of moss; that it was alert and listening intently with pricked-up ears." Arethusa blossoms from May to August in bogs and peaty meadows throughout the northeastern United States and Canada. Its abundance in any location varies tremendously from year to year: in one northern Michigan bog, the number of plants ranged from twelve to more than a thousand. Arethusa was named for a nymph who, in Greek mythology, was transformed into a spring so that she could escape an amorous god.* (John Shaw)

140 *overleaf. The showy ray florets of most flower heads in the tickseed genus are sun-yellow, but those of swamp coreopsis* (Coreopsis nudata), *found in ponds and wet pinelands of Florida and Georgia, are rose-purple. Tickseed seeds do indeed look like ticks.* (Wendell D. Metzen)

142. *An ant is being lured to its death in the inch-high jug of an Albany pitcher plant* (Cephalotus follicularis), *found in moist coastal forests of southwestern Australia. Attracted by bright colors and aroma, the ant was led to the rim by a vertical "fence." Comb-like teeth on the rim, permit easy entrance to the pitcher—but preclude exit—and the ant will soon slide into and drown in a well of powerful digestive fluid. Of 250,000 species of flowering plants on Earth, some 400 are carnivorous. Growing in nutrient-poor habitats, they have evolved their meat-eating techniques to provide vital phosphorous and nitrogen.* (Michael Morcombe)

Fly Traps and Other Insect Eaters

My wife and I were plant hunting in second-growth tropical forest less than an hour from Singapore. Before us, covering the ground in dense masses, were dozens of small green urns of demitasse size, each seemingly with its own tiny teacup-like handle and brimful of rainwater. Beyond them, hanging here and there in the undergrowth, were other, much larger vases of several shapes—some green, some with maroonish mottling and all with jaunty parasol-like caps held high over the liquid-filled containers. These proved to be suspended by strong, green cord-like tendrils originating from the leaf tips of several sprawling vines. We had found a trio of equatorial pitcher plants (*Nepenthes*), the best-known insectivorous plants in the Old World tropics.

Insects that eat plants are commonplace, but plants that eat insects always evoke wonder. Unlike other plants, they supplement their normal photosynthetic diet with the protein-rich "meat" of insects. Most insectivorous wildflowers live on poor soil. Thus it seems that the unique insect-eating habit has developed as an evolutionary response to survive in habitat niches where few flowering plants can grow. Such nitrogen-deficient sites occur around the world.

Because wildflowers are stationary and cannot pursue insects, they have had to devise methods to lure and then trap their unsuspecting prey. These devices include passive pitfalls, flypaper-like lures and complex spring-like traps. All these traps and snares are simply ordinary green leaves that have been remarkably transformed. Even their color is often changed, usually to shades of red, producing a flower-like lure to decoy the prey. Most transformed leaves serve both as traps and as digestive "stomachs." The same kinds of glands that secrete nectar in flowers are utilized to secrete enzymes for decomposing animal foodstuffs, permitting them to be absorbed into the leaf. Pitcher-producing plants, with curious watery pitfalls constructed of tubular water-holding leaves, are among the larger insect-catching wildflowers. Curiously, three families of wildflowers in three different areas of the world have independently "discovered" the usefulness of this type of trap, and their three kinds of pitchers are an interesting example of convergent evolution. In the closely related pitcher plants of the New World— the *Sarracenia* of eastern North America, the California pitcher plant (*Darlingtonia californica*) and the sun pitchers (*Heliamphora*) of the "Lost World" mountains of the Guianas—the whole leaf is transformed into a pitcher trap. In the climbing tropical pitcher plants (*Nepenthes*) of Indomalaysia, many leaves are normal, and even in the traps the lower portion remains leaf-like while the tendril tip changes into the often very large pendant pitcher. In the West Australian pitcher plant (*Cephalotus*), some leaves remain ordinary leaves.

Without exception, pitcher plants are alluring, often with brightly colored blotches or stripes of red on a greenish or yellowish background. Thus each trap resembles a flower, with the aroma of nectar, the bait, produced by special glands located at the pitcher's mouth. Encircling the mouth of most pitcher traps is a slick rim that acts as a slide and as a barrier to prevent insects from escaping once they have entered the trap. In the Australian species, the pitcher rim is guarded by a formidable barricade of teeth that permit entrance but not exit. Within the pitcher, just below the rim, is another slide zone, a final shoot-the-chute that delivers the insect to certain death in the smelly well. Glands cover this slide zone and secrete digestive enzymes that accumulate in the liquid below. Any prey that falls into the cistern soon drowns and is decomposed rapidly, yielding nitrogenous food for the plant. Sundews and butterworts are insectivorous plants that trap insects on sticky living flypaper. Their prey are small flying insects, gnats, midges and mosquitoes, which can be accommodated on the small leaves of these rather insignificant plants. There are hundreds of kinds of true sundews (*Drosera*), which occur in all major habitats around the world, from arctic tundra to tropical mountains. Most sundews are small rosette plants, easily overlooked. The Portuguese sundew (*Drosophyllum lusitanicum*) may be five feet tall, large enough to have been collected at one time or another by Portuguese countryfolk and hung in their cottages as natural flypaper. In southwest Australia, world center of sundews, the variety of growth form and flower color of these plants is unbelievable. One memorable day on the damp sand plains north of Perth, I found sundews with big leaf rosettes four inches in diameter; others had erect, foot-long grass-like leaves; one species had knee-high branching stems; and there were even climbing sundews, with yellow, red or pink flowers up to the height of a man.

Sundews frequent wet sites like damp sands or the sphagnum bogs of the north. Although their leaves may vary in shape, they share a common insect-catching function. Those of the ordinary round-leaf sundew are typical. Its spoon-shaped blades are borne on tapering petioles. Scores of reddish tentacle-like glands cover the margins and upper surfaces of the leaves. The rounded tips of the glands secrete globules of a sticky fluid, the "sun dew."

Whatever the lure, when an insect lands on a sundew leaf it is promptly snared. Initially mired in a mucilage, the prey struggles, thereby stimulating secretion of adhesive by the glands. So the greater the struggle, the more hopeless the insect's situation. The moment an insect lands, all tentacles bend toward the prey, enveloping it. Then smaller glands on the leaf's surface start to digest the prey. The trapping process requires but a few minutes, while several days are required to consume the meal, after which the leaf reopens.

Butterworts (*Pinguicula*) are nearly as widespread as sundews. Their name comes from the buttery "feel" of their yellow-green tongue-shaped leaves. Each leaf is covered with hundreds of short glands which secrete the insect-snaring mucilage. Other, shorter glands are also present on the leaf and secrete the digestive enzymes. In butterworts the leaf margins are the sensitive structures, and when an insect is caught, the leaf edges roll inward, enveloping the prey and thus permitting digestion and absorption.

Remarkable though they are, flypaper traps fail to have the visual impact of the insect-catching apparatus of the

145

Venus flytrap (*Dionaea muscipula*). Few wildflowers occupy such a restricted range—only a few score square miles on the coastal plain of the Carolinas—yet the species is locally common. On one of my visits summer was just starting, and Venus flytraps were spreading their colorful rosettes of bright red leaves upon the sandy ground.

Several flytraps were in flower. The white blossoms on their foot-high stalks proclaimed their kinship with the sundew family. Flytrap leaves, however, are not like those of sundew. Each rounded blade is neatly hinged at the midrib, forming identical halves, which close to make a remarkable hinged leaf trap. On the leaf margins are comb-like teeth which interlock like fingers when the trap closes. The bright red color of the leaf surface marks the location of thousands of closely packed short-stalked glands, which have secretory and absorbing functions. Rising from the surface of each lobe are three stiff hairs, the sensitive triggers of the trap. They react only when one or more are touched twice. To observe the trap action, I caught an ant and placed it on the edge of a flytrap leaf. Crawling over the colored surface, it promptly touched one trigger hair: there was no action. But a moment later another hair was touched. Zip! Within a second or two, the trap closed quickly, pressing the ant tightly against the glandular surface and thus starting the enzyme-secreting process.

Not all insectivorous plants are terrestrial; some are aquatic. The bladderworts (*Utricularia*), cousins of the butterworts, inhabit ponds and slow-moving streams. A typical bladderwort lives a floating existence, with only its small snapdragon-like flowers raised above the water. Its leaves are finely divided, but attached to them are minute bladders, each a tiny active trap fully as complex as a Venus flytrap. But these submerged traps feed largely on aquatic animals. Each bladder trap is a hollow, transparent, pear-shaped structure with a circular trapdoor guarded by long branching hairs and fitted with a watertight groove. The whole interior surface of the bladder is lined with tiny hairs that continually absorb water within the structure. Thus a negative pressure is built up inside the bladder. When a small swimming creature touches a trigger hair, the one-way door trips, and the prey, carried in by a current of water, is swept to its doom.

147. *Endemic to bogs in California and Oregon, the California pitcher plant (Darlingtonia californica) is more commonly known as the cobra lily, for its hooded leaves give it the appearance of a cobra about to strike. Land development and poaching for the houseplant trade have made it a rare find in many of its haunts. This pitcher plant provides flying insects with a convenient landing ramp, a fishtail-shaped appendage in front of its opening. Translucent spots in the walls let sunlight reach the interior of the pitcher, so that insects will not be discouraged from entering a dark tube. Once inside, they try to escape through these "windows," eventually tire, and fall to the bottom. Most carnivorous plants secrete enzymes that digest victims; when an animal is decomposed, the plant absorbs the end products through its outer cell walls. In the cobra lily, digestive action is due to bacterial flora in the pitcher's pool. Pitchers last for several months, and the old ones often are nearly filled with decaying remains of their catch. (Gary Braasch)*

148. *Animals as large as mice have fallen, perhaps by chance, into the beautifully sculpted pitchers of the genus* Nepenthes, *which hang on long leaf stalks twined around tree branches in the forests of tropical Asia. Insects are lured by nectar-secreting glands on the lip of the vase, tumble in, and are then unable to scale the slippery inner wall. Yet other creatures, including several kinds of insects and a spider, are able to live and breed inside the pitcher, unharmed by the quart or more of digestive fluid it holds and able to climb out at will, despite a battery of barriers that frustrate the escape of most of the plant's victims.* (Oxford Scientific Films)

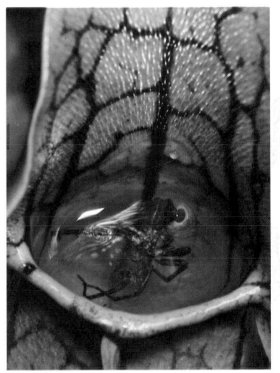

149. *A spider is being consumed by a northern pitcher plant* (Sarracenia purpurea), *trademark of sphagnum bogs from Canada to Florida. Insects follow a trail of nectar up the outside of the tube; once they are past the rim, their retreat is thwarted by dense, stiff, downward-pointing hairs. Because these pitchers lack hoods, enzymes in their pools become diluted by rainwater, so the plants also depend on bacteria to help digest their prey. The usual victims are ants, beetles, and flies; but the list also includes toads, frogs, and centipedes. Tree frogs, however, cling to the inner walls, snatching insects before they fall to the bottom; and the enzyme pools often are populated with mosquito larvae.* (Betty Randall; Dwight R. Kuhn)

150 *all. Though no larger than the nail on one's little finger, this single leaf of a round-leaved sundew (*Drosera rotundifolia*) in a Michigan sphagnum bog is covered with more than 200 bright-red tentacles. Each bears a globule of sticky, transparent fluid exuded by the gland at the tip of the stalk. An ant of the genus* Lasius, *which gathers honeydew excreted by such insects as aphids and leafhoppers, has been attracted by these shining diamonds of "sundew." Snared by the adhesive drops, the struggling ant will be quickly enveloped by bending tentacles that push the captive to the center of the leaf. The ant will be dead in a few minutes; but the trap will not be "reset" for several days, until digestion is finished.* (above, John Shaw; *center and below,* Oxford Scientific Films)

151. *Native to the eastern coastal plains of the United States, thread-leaved sundew (*Drosera filiformis*) unfurls like the fiddlehead of a fern. Though most sundews are diminutive, only an inch or two across, they occasionally form spectacular, glistening swards. An estimated 6 million cabbage butterflies, migrating across the English Channel from the Continent, were trapped one morning when they settled on a two-acre stand of sundews on the Norfolk coast. Four to seven butterflies were caught by each of several hundred thousand plants. Charles Darwin was the first to show the importance of animal prey as a nutrition supplement for carnivorous plants. The sundews he fed artificially by sticking insects to their leaves were more vigorous and produced more flowers and seeds than those from which prey was withheld.* (Ruth Allen)

152 *overleaf. The only continent lacking sundews is Antarctica, and Australia has more species than any other, among them the beautiful rosettes of spoon-leaved sundew (*Drosera spathulata*). That carnivorous plants can cope with environmental extremes in amply demonstrated by sundews in northwest Australia: during the wet winter months, their growing season, they are exposed to near-freezing temperatures at night; during the dry summer months they lie dormant while daytime temperatures may exceed 120°F.* (Ernst Müller)

154. *Like out-of-place snapdragons, the buttery flowers of swollen bladderwort* (Utricularia inflata) *rise in spectacular profusion above a shallow pond in Georgia's Okefenokee Swamp. They are supported by a wheel of inflated leafstalks that float just below the surface. While bladderworts depend on flying insects to pollinate their flowers, the seeds that form will be nourished by the bodies of aquatic insects trapped and digested underwater. Some 250 kinds of bladderworts thrive in sluggish waters from the Arctic to the tropics; a few species live an epiphytic existence in the wet moss of cloud forests.* (Wendell D. Metzen)

155 *both. Underwater photographs show the feeding technique of common bladderwort* (Utricularia vulgaris), *found in ponds across the north temperate and boreal zones. Scattered throughout its floating, filament-like leaves are elastic-walled bladders no more than a quarter-inch wide. These suction traps, when set, are flat and nearly empty of water. The mouth of the bladder is fringed with trigger hairs: when a passing animal—here a mosquito larva—touches the trigger, a flap of tissue forming the bladder door springs open, water rushes in, and the prey is carried with it. The water is removed through absorbent cells, the animal quickly disintegrates and is digested within an hour or two, and the trap resets itself. Bladderwort prey ranges in size from diatoms to small crustaceans such as daphnia to newly hatched fish. Larger victims may be caught by only part of their bodies.* (Oxford Scientific Films)

156 and **157.** *Though widely cultivated,* Venus flytrap (Dionaea muscipula) *is endemic only to a few wet sandy habitats on the coastal plain of North and South Carolina. Two identical lobes edged with comblike teeth and hinged at the middle form the flytrap's leaf blade, which may be two inches wide when open. In bright sunlight, thousands of enzyme-secreting glands give the leaf the appearance of a bright-red flower. This and sugary nectar attract a fly; investigating the nectar source, the insect brushes against stiff trigger hairs, and in a second or two the leaf slams shut, its teeth interlocking like fingers. A Venus flytrap requires several days to digest its meal. Eventually the leaf reopens, and the desiccated remains of its victim blow away in the wind. A single leaf shrivels and dies after feeding three times. (Robert W. Mitchell)*

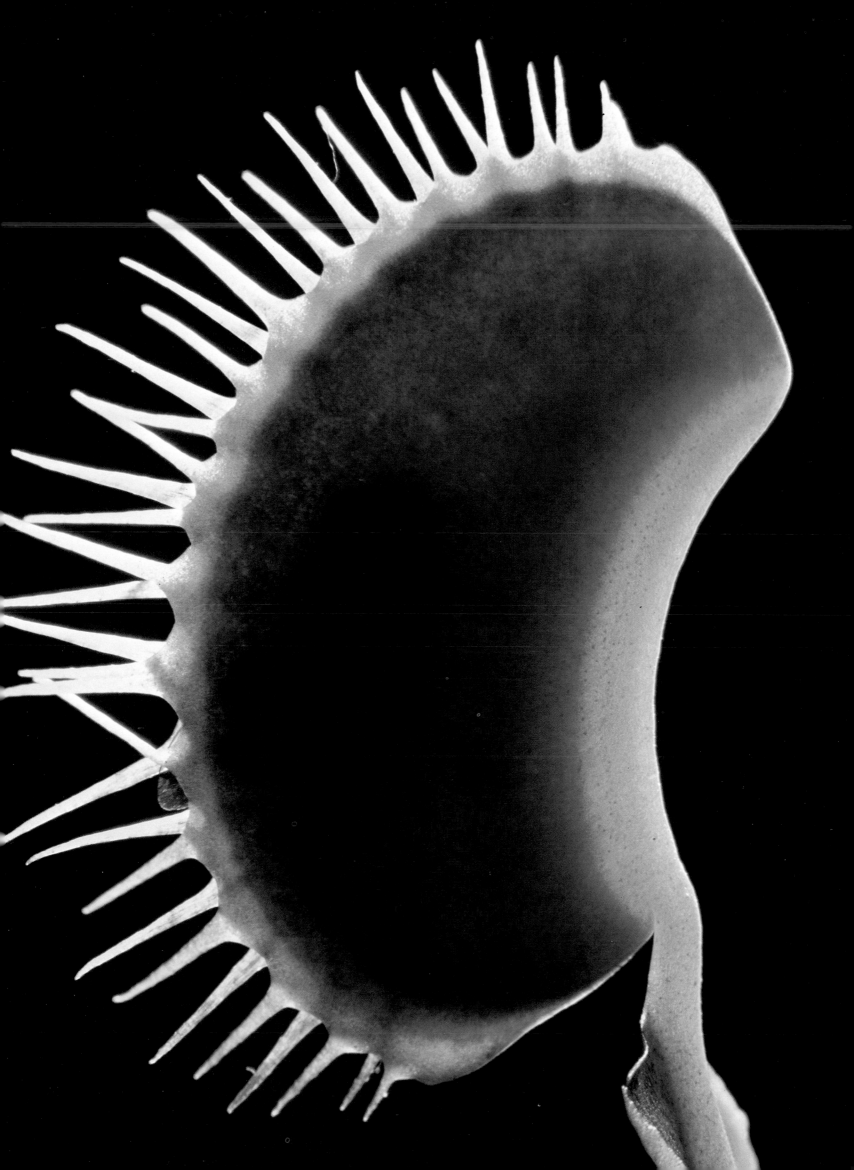

158. *In early April, the lacy leaves of Dutchman's-breeches* (Dicentra cucullaria) *occupy every available niche among the tree roots that anchor a Virginia hillside. Close examination of the delicate half-inch-long flowers explains their quaint name: they do indeed suggest white pantaloons with yellow-belted waists swaying on a clothesline. As many as ten pairs may hang from a single stem; their "legs" or spurs are modified petals containing nectar glands. Queen bumblebees, which alone survive the winter to start new colonies in the spring, pollinate Dutchman's-breeches by probing the spurs with their long tongues.* (William A. Bake)

Flowers on the Forest Floor

The delicate pink blossoms of fringe bells (*Shortia soldanelloides*) are one of the loveliest wildflowers to appear in spring in the montane deciduous forests of Japan. Fringe bells and a cousin species, Nippon bells (*Shortia uniflora*), are among Japan's most unusual wildflowers. For these two little vernal herbs have only a single sister species and, curiously, that relative is the rare little Oconee bells (*Shortia galacifolia*), which lives thousands of miles away in the deciduous forests of western North Carolina. This wildflower trio obviously had the same ancestors, but how did they come to live so far apart?

Actually there is a small group of woodland wildflowers having similar pairs of closely related species, one occurring only in eastern Asia and the other in eastern North America. Trailing arbutus (*Epigaea repens*), brook saxifrage (*Saxifraga*), witch hazel (*Hamamelis*), leucothoë (*Leucothoe*), blue cohosh (*Caulophyllum thalictroides*) and the crane-fly orchid (*Tipularia*) are among wildflowers of eastern North America with cousins in Japan. These Asiatic and American plants are all denizens of the deciduous forest habitat, and it is in the past history of this summer-green forest that an explanation is found for the curious geographical discontinuities of some of its wildflowers.

Most of the world's forests are evergreen or essentially so. This includes the great belt of needle-leaved conifers that makes up the boreal taiga, the humid tropical rain forests with their countless kinds of broad-leaved evergreen trees, and the hard-leaved trees of the widely dispersed woodlands of Mediterranean-type climes. Deciduous forest, on the other hand, is the common forest habitat of much of the north temperate belt, where a cold but not severe winter season annually interrupts the growth of plants. The deciduous habit, the annual fall of leaves, is a modification developed to conserve loss of moisture during periods of drought. The broad leaves of deciduous trees, unlike the narrow wax-covered leaves of the conifers, normally transpire, or give off, large amounts of water daily. By dropping all their leaves at the onset of drought, trees can reduce life processes and thus survive the desert-like conditions of winter, when water is frozen.

Deciduous forests were the original woodlands of temperate Europe and of eastern Asia, including much of China, Korea and Japan, as well as those parts of eastern North America that lie south of the taiga and east of the prairies and plains. Unlike conifer forests, deciduous forests do not form a continuous belt around the globe, but rather are isolated from each other by other major habitats or by great seas.

The world's deciduous forests have not always been so isolated. Scientists know that in early Cenozoic times—about 60 million years ago—climatic conditions were milder and more favorable for development of such woodlands. Thus a wide, continuous belt of deciduous forest once covered all of the northern continents. Within that vast habitat thrived the ancestors of many temperate forest wildflowers, including those of Japan's fringe bells.

In those days, wildflowers of the deciduous forest had a much wider distribution. Subsequently there was an onset of glaciation. The climate changed, bringing about more rigorous conditions for plant growth. One result was that the former continuous belt of deciduous forest contracted and was pinched into parts that are now widely separated. The worldwide change to drier conditions brought other habitats into the picture. In Eurasia and North America were conifer forests, grasslands and desert. All these newer habitats survive today. Thus the once-continuous deciduous forest of eastern Asia and North America, as elsewhere, was reduced to its present, much smaller area. Many wildflowers that once had a more extensive range were split into separate populations. Inevitably the long isolation brought gradual evolutionary changes that have made certain Oriental wildflowers different from their American cousins.

The fact that the world's temperate deciduous forests were once one great continuous woodland also explains why today the majority of the trees and wildflowers of Europe, Japan and the eastern United States are so similar. Thus a traveler from one of these regions to any of the others soon sees that the forest trees, whether oaks, beeches, birches, lindens, chestnuts, elms, hickories, ashes or maples, are similar to those in his homeland. Also familiar are the violets, wood anemones, buttercups and Lady's-slipper—common in all these deciduous forests.

Of course, there are some differences: European forest wildflowers do not have the variety found in Japan or eastern America. But America lacks the primroses (*Primula*), pheasant eyes (*Adonis annua*), and lilies-of-the-valley (*Convallaria*) of Europe and Asia. Each of the world's major deciduous forests has its unique wildflowers. America has bloodroots (*Sanguinaria canadensis*) and *Phacelia*, Europe its snowdrops (*Galanthus*), and Japan the toad lily (*Tricyrtis*), kirenge-shoma (*Kirenge-shoma*) and shirane-aoi (*Glaucidium*).

Spring is the main season for wildflowers in all deciduous forests. Only then is the ground warm enough to encourage the rapid growth of herbaceous plants. For several weeks the awakening trees will be leafless, and while they are half-dormant nature produces a garden of woodland flowers. But before the main show there is a pre-spring flowering. Indeed, where climates are mild, some wildflowers may bloom during winter itself. These are often woodland shrubs, such as camellias and witch hazels, which in some species bloom from late autumn to early spring.

In March or April, when treetop temperatures are still cool, keeping the swollen buds of the trees from opening, the very first wildflowers appear. Then the forest floor has a microclimate of its own. The ground, protected by the still leafless trees and shrubs, is relatively warm and at its very sunniest. Since annuals cannot mature in the short time available before the leaves emerge and shade the ground, these early wildflowers are all perennials. Their underground parts are usually thickened storage structures: rhizomes, corms, or bulbs, packed with food manufactured by leaves of the preceding season. Thus these pre-vernal plants can expand their flower buds immediately, for they too were pre-formed a year earlier, so as to be ready for the first warm days of spring. In all temperate woodlands the widespread *Hepatica*, in one species or another, is among the first of the pre-vernal flowers. It shares early spring honors in Europe with snow-drops and snowflakes (*Leucojum*), winter aconite (*Eranthis hyemalis*) and lesser celandine (*Ranunculus ficaria*). The golden Amur adonis (*Adonis amurensis*) is companion of *Hepatica* in Japan, while in America there are bloodroots, spring beauties (*Claytonia*) and Dutchman's-breeches (*Dicentra*).

The major show of flowers accompanies the opening of the buds of the forest trees; the latter bloom, though inconspicuously, with the showy display of herbs at their feet. As the young tree leaves expand, a green curtain is gradually drawn over the forest floor, reducing sunlight and fully shading the ground. Leaves that are large in proportion to the plant, or that form vast mosaics, are modifications developed by spring wildflowers enabling them to take best advantage of the steadily diminishing light.

Woodland wildflowers have short life-cycles carefully controlled by the alarm clock of day length to match those few days in spring when sunlight is adequate. During that short period they flower, leaf and produce seed, and by late spring have stored food and formed buds in readiness for another year. When the ground is shaded, many have become dormant and another group takes over the stage.

The deciduous forest continues to play host to a succession of wildflowers, each occupying its own place and giving with the others a conspicuous seasonal aspect to the forest floor. But by mid-June the main show is over, for there is no longer enough sun to produce sheets of flowers. The later blooms are more likely to be found in forest clearings or along its borders. Shrubs such as rhododendrons and other heathy plants often flower in early summer, as do the woodland hydrangeas (*Hydrangea*) of Japan. Among herbaceous perennials are wood lilies (*Lilium*), goatsbeard (*Aruncus*), plantain lilies (*Hosta*) and bee balm (*Monarda didyma*), while rank-growing annuals such as touch-me-nots (*Impatiens*) bloom in late summer.

Whereas spring is dominated by the members of the lily, rose, buttercup and mustard families, late summer and fall belong to the asters (*Aster*), whose family representatives also include goldenrod (*Solidago*), ironweed (*Vernonia*), snakeroot (*Eupatorium*) and tsuwabuki (*Farfugium*).

163. *The first color of spring in the beech-maple woodlands of southern Michigan is likely to be the fragile little flowers of round-lobed hepatica (Hepatica americana) waving vigorously in the breeze. Hepatica has no petals; its flowers, which attract early-emerging bees and flies, are formed by a ring of pigmented sepals that may be white, pink, blue, or lavender. Hidden beneath the forest litter are last year's hepatica leaves; new ones appear after the flowers. Because their three lobes resemble the shape of a human liver, hepatica was given a name that, in Greek and Latin, means "liver-like"; and the plant was used in herbal concoctions to treat liver ailments. (John Shaw)*
164 overleaf. *Bur marigolds (Bidens sp.) are also known as sticktights and beggar-ticks because their seed-like fruits are armed with tiny hooks that grab the clothing or fur of any human or animal brushing past. Like these dew-covered flowers beside a Michigan lake, most species in this large worldwide genus grow in damp or swampy places. (John Shaw)*

166. *A relative of the daffodil and narcissus, spring snowflake* ° (Leucojum vernum) *is among the first showy flowers of the year in the wet woods, copses, a[nd me]adows of northern Europe. Pushing through lingering snow in February and March, it unveils a single green-tipped white bell on each leafless stem.* (Hans Reinhard/Bruce Coleman, Ltd.)

167. *Wildflower enthusiasts can find a wealth of trilliums—eight species and two varieties—in the Great Smoky Mountains of North Carolina and Tennessee. Prominent in spring forests up to elevations of 5,000 feet are the lemon-scented flowers and mottled leaves of yellow trillium* (Trillium luteum). *The name "trillium" comes from the Latin word for "three": all parts of the plant occur in groups of three—three leaves and a single flower with three green sepals, three petals, six stamens in two whorls of three, and a three-chambered pistil with three stigmas.* (Sonja Bullaty)

168 *overleaf. Tall grasses cannot hide the fiery beauty of fire pink* (Silene virginica) *along a trail in the southern Appalachians. A member of the vast pink family, this is a relative of such popular cultivated flowers as carnations, sweet William, and baby's breath.* (Sonja Bullaty)

170 *second overleaf. The Turk's-cap lily* (Lilium superbum) *certainly is, as its botanical name states, a superb lily. The largest and most spectacular of North America's twenty native lilies, it grows as tall as eight feet, with as many as forty flowers on a single stem. The three sepals and three petals of the spotted orange-red flower are curled back sharply to reveal six long pollen-forming stamens and an equally long three-lobed pistil, the female seed-forming organ.* (Les Line)

172. *In summer the large, upright flowers of nettle-leaved bellflower* (Campanula trachelium) *decorate the woodlands and hedges of northern Europe. The bellflower family has some 2,000 species growing throughout the world; best-known is the common harebell* (Campanula rotundifolia), *found on meadows and heaths around the northern hemisphere.* (Ingmar Holmåsen)

173. *Bluets, Quaker-ladies, innocence, angel eyes — this tiny spring wildflower* (Houstonia caerulea) *of North American woodlands has at least a dozen names in the English language. Its yellow-eyed blossoms, with petals ranging in color from rich lavender to blue-white, are only one-third of an inch across and are borne on delicate stems hardly taller than a blade of lawn grass.* (Sonja Bullaty)

175. *One of the least imposing wild-flowers of the North American springtime, miterwort (*Mitella diphylla) *needs intimate examination to be fully appreciated. Under a hand lens, its quarter-inch flowers resemble beautifully sculpted snowflakes, with five fringed petals surrounding ten stamens. The common name comes from the tiny seedpod, shaped like the ornamented miter (or cap) of a bishop.* (Larry West)

174. *The creamy yellow bells of wild oats (*Uvularia sessifolia), *also known as merrybells and strawbells, appear in mid-spring in the deciduous woodlands of eastern North America. Its young shoots can be boiled and served like asparagus or chopped into a salad.* (Farrell Grehan)

176 *overleaf. The flowers of red trillium (*Trillium erectum), *it is said, greet migrant American robins returning in spring to northern states; thus one of its several common names, wake-robin. In truth, robins arrive on the scene many days before the maroon-red flowers open in damp woodlands. Red trillium is also called stinking Benjamin because of its foul scent, which lures carrion flies to the task of pollination.* (James Cunningham)

178. *Oconee bells* (Shortia galacifolia), *an early spring flower found only in the Appalachian Mountains of North and South Carolina, played hide-and-seek with science for nearly a century. A sample was first collected in 1788 by the French botanist and traveler André Michaux; but his specimen lay unknown and undescribed in a Paris herbarium until the early 1840s, when it was discovered by the great American botanist Asa Gray. Then, however, it took collectors until 1877 to find another living plant—even though it is not a particularly rare species.* (Patrick W. Grace)

178. *The underground stem of bloodroot* (Sanguinaria canadensis) *oozes orange-red juice that American Indians used as a dye, as warpaint, to repel biting insects, and to treat rheumatism. Bloodroot loves the leafless sunshine of early-spring woodlands; its emerging bud is wrapped in a large leaf that does not expand fully until the flower opens.* (C. W. Perkins)

178. *Gaywings* (Polygala paucifolia), *sometimes mistaken by beginners for an orchid, is often called "flowering wintergreen" because of its evergreen leaves but it belongs to the milkwort family. In the Old World, milkwort once was given to cows and nursing mothers to increase lactation. The one-inch flowers of gaywings bloom from May to July in rich forests from Quebec south to the mountains of Georgia.* (John Shaw)

179. *On a Swedish island in the Baltic Sea, bloody cranesbill* (Geranium sanguineum) *unveils its rare pink-flowered form. Normally this species, which thrives in limestone habitats, has a crimson-purple flower.* (Ingmar Holmåsen)

179. *The pretty blue blossoms of dayflower* (Commelina *sp.*) *open early in the morning and disappear by afternoon. Usually there are two showy petals and a third petal that is white and insignificant. Linnaeus whimsically named the genus after three brothers of his acquaintance: two were noted botanists, the third accomplished nothing of scientific worth. Asiatic dayflower has become a widespread, albeit colorful weed in North America.* (Robert W. Mitchell)

179. *Four delicately fringed petals swirling out from a deep corolla tube instantly identify the beautiful fringed gentian* (Gentiana crinita) *of damp, secluded sites in North American woods and meadows. The flower's name honors Gentius, a Balkan king of ancient times who promoted medicinal use of gentian roots. Among the last flowering plants of autumn, gentians appear from September to November. The fringe, which collapses under the slightest weight, prevents crawling insects from entering the flower.* (Dorothy M. Richards)

180 *overleaf. Touch the leaves of sensitive brier (*Schrankia *sp.), and they will immediately fold up. This sprawling, prickly plant displays its rose-pink flower heads from May to September in dry, sandy woodlands across the southern states. It is one of the few North American representatives of the mimosa family.* (Robert W. Mitchell)

182. *"It is like an enormous, brilliantly red asparagus stalk,"* one scientist wrote of the snow plant (Sarcodes sanguinea), *which appears from May to July in mountain forests from Baja California north to Oregon. Said one wildflower guidebook, "A glance at the illustration makes description unnecessary; no other plant of our forests can be confused with this." The foot-high raceme of fleshy flowers pushes through the forest litter of needles and bark while snow still lingers on shady slopes—thus its name. Snow plant is a saprophyte: it has no green, foodmaking leaves and depends on a fungus intermediary to obtain sustenance from other plants.* (Ed Cooper)

Under the Firs and Spruces

My first extensive field trip as a young botanist was to the coniferous forests of Canada. Perhaps that is why I have always thought of the North Woods by that romantic old name, "land of the pointed firs." In retrospect one forgets the myriad mosquitoes and blackflies of that long-ago summer, the difficulty of forcing one's way through the barrier of conifer branches or taking circuitous paths to avoid floundering in bog or muskeg. Instead one savors memories of the pervasive peace of the forest, the plaintive calls of the nesting whitethroats, the fragrance of balsam and the feel of the mossy forest floor. And of course, there were the wildflowers typical of northern coniferous forests the world round—wood sorrel (*Oxalis*) and constellations of goldthread (*Coptis*) and starflowers (*Trientalis*), creeping snowberry (*Gaultheria hispidula*) and the delicate trailing twinflower (*Linnaea borealis*).

That bit of Canadian North Woods is typical of a great belt of pointed-tipped conifers spanning the northernmost continental masses of North America and Eurasia, just south of the tundra. It cloaks hundreds of thousands of square miles of Alaska, Canada, Scandinavia and the Soviet Union and remains today one of the world's most extensive forest habitats. The Russians call this forest "taiga." Its other names, boreal forest or northern coniferous forest, are more meaningful.

Patches of coniferous forest, sometimes somewhat modified in aspect, also extend southward in subalpine or other sites on mountains, occupying positions just below the alpine tundra. In North America such extensions are seen on the upper slopes of the Cascades and Rockies, as well as the Appalachians, and the same is seen on the mountain ranges of Europe and Asia. But upon moving from north to south, one must climb progressively higher on the mountains to reach the zone of conifers. In eastern Kamchatka conifers occur at sea level, in central Japan at nearly 5,000 feet, while hundreds of miles farther south on Taiwan (Tropic of Cancer) the lowermost reaches of conifer forest are at over 8,000 feet.

Lying just south of the treeless tundra, the boreal forest has a continental climate almost as severe as the Arctic, with long, cold, snow-filled winters alternating with rather short, cool, moist summers. However, not even the tundra matches the sub-zero temperatures of the Siberian taiga, where the thermometer has plunged to −93.6°F (−69.8°C), the lowest recorded temperature on earth. But the forests hold the snow, and snow cover is a protection to their wildflowers. Even so, plants of the taiga are among the world's hardiest.

Wherever coniferous forest occurs, whether it be in Scandinavia, Siberia or Saskatchewan, it has a similar look. In North America the dominant conifers are the white and black spruces, balsam fir and tamarack, whereas in the Soviet Union there are Siberian spruce, Siberian fir and Siberian larch. Similar differences occur in the southern extensions of coniferous forest.

The needle "leaves" of the conifers, which offer a minimal surface for water loss (transpiration), enable the trees of the taiga to remain evergreen year-round, with the ability to manufacture food whenever conditions permit, even during the drought of winter. It is almost as if the trees, in a habitat where the growing season is all too short, have found it essential to keep their leaves; for among the conifers, only larches are deciduous, dropping their needles in autumn after a blaze of yellow color.

Besides the cold and desiccation winter brings, there are other hazards for boreal forest plants. Most of the land has been scoured by glaciation. Soils, acid and infertile, usually are thin over the underlying rock; ponds and bogs dot the landscape, all in one stage or another of being transformed to muskeg through gradual accumulation of peat. Though forming dense forest, the trees are seldom as much as 50 feet tall. Where conditions are most severe—at the forest-tundra boundary or at the treeline on mountains—trees show the effects of their grim struggle with the elements. They become contorted mats, aptly described in German as *krummholz*, meaning "twisted wood."

As evergreens, the trees not only shade the ground in all seasons but also continually drop resinous decay-resistant needles to the forest floor, building humus that is distinctly acidic. As any gardener knows, the combination of shade and acid soil is not conducive to growing a great variety of plants. Thus coniferous forest wildflowers are nowhere near as numerous as those found in deciduous forests. Most are perennials; as in the tundra, summers are too short to permit the establishment of annual wildflowers. That there is any variety is due primarily to the prevalence of bogs, muskeg and rock outcrops, which allow the growth of some sun-loving wildflowers.

The commonest woodland wildflowers include those mentioned earlier, as well as the little dwarf cornel or bunchberry (*Cornus canadensis*) and the Canada mayflower or false lily-of-the-valley (*Maianthemum canadensis*). The little bunchberry is attractive in flower as well as in fruit. Its "flowers" resemble in miniature those of its close cousins, the flowering dogwoods (*Cornus florida*) of the temperate forests of North America and eastern Asia. Flowering shrubs also occur, often at breaks in the forest. Among the more showy species are several viburnums (including hobblebush), wild currants (*Ribes*), salmonberry (*Rubus spectabilis*) and salal (*Gaultheria shallon*) of the coastal coniferous forests of America's Pacific Northwest.

The pervasive acidity of forest soils encourages many "heathy" wildflowers. These include not only members of the heath family but also its allies, the pyrolas (*Pyrola*) and wintergreens (*Chimaphila*), as well as a sprinkling of orchids. Each of these wildflower groups has representatives in both shady and sunny sites. Calypso, or fairy slipper (*Calypso bulbosa*), is the most beautiful of the taiga orchids and occurs in boreal forests around the world, sharing shaded glades with some less showy cousins—twayblades (*Liparis*), coral-roots (*Corallorhiza*) and rattlesnake plantains (*Goodyera*).

Orchids and heaths can thrive in acid soils because of

an unusual relationship between the roots of these plants and certain specialized soil fungi called mycorrhizae—literally, "fungus roots." Mycorrhizae live symbiotically upon or within the feeding roots of the host plant, decomposing and feeding upon organic material in the soil and sharing the food with their host in return for a "home." Thus it is the fungi, not the hosts, that require the acidic conditions in which to prosper. All orchids are especially dependent on the help of mycorrhizal fungi. Their dust-like seeds lack the stored food normally supplied in most seeds to sustain the embryo plant during its germination. A seed of an orchid will not germinate unless first penetrated by mycorrhizae. If that happens, the tissues of the fungus help nourish the infant seedling until the first green leaves take over the task of food-making. Some plants have apparently lived so long and intimately with mycorrhizae that they have lost their power to develop green food-making leaves. These are the saprophytes, which depend on their fungus partners for all organic foodstuffs. Since they lack green leaves, such plants easily thrive without sunlight. The northern coralroot orchid (*Corallorhiza trifida*) and the pale, waxy Indian pipe (*Monotropa uniflora*) are examples of saprophytes that are familiar denizens of coniferous forests.

Quite a different group of boreal wildflowers inhabits the treeless areas of the taiga, especially the bogs and muskeg, where conditions are too wet for the growth of conifers. In time, trees eventually reclaim these wet areas, filling them in with forest. This is a long cycle. Bog mosses and flowering plants create a peaty land-fill which gradually converts open water first to bog and then to firmer soil. Spruce and larch eventually become established. Bogs and muskeg, prior to the coming of trees, have their own special wildflowers. Shrubby members of the heath family are especially prominent—bog rosemary (*Andromeda glaucophylla*), pale laurel (*Kalmia polifolia*), blueberries (*Vaccinium*) and bog cranberry (*Vaccinium oxycoccus*).

Delicate northern fringed orchids (*Habenaria*) and cotton grasses (*Eriophorum*) often dot the bogs in summer, while hidden in the sphagnum are the insect-catching sundew (*Drosera*) and northern pitcher plant (*Sarracenia purpurea*). Thus, although the world's conifer forests do not have a great number of wildflowers, those they do have seem doubly attractive.

187. *The waxy, ghost-white flowers of Indian pipe (Monotropa uniflora) are a familiar sight to summer hikers in northern forests near the pole. Like snow plant, this is a saprophyte; it was long believed that such flowers, lacking chlorophyll, lived on dead organic matter. Botanists now know that these plants obtain food from tiny soil fungi, or mycorrhizae, which form a living bridge to the roots of photosynthetic plants. The single bell-shaped flower of Indian pipe belies its membership in the heath family; it is thus a relative of blueberries and cranberries, of azaleas and rhododendrons. As the seed forms, the nodding flower, which sometimes is pink, turns upright. American Indians treated eye problems with juice from its crushed stems.* (C. W. Perkins)

189. *Twinflower* (Linnaea borealis) *is a delicate, fragrant evergreen of cold northern forests and bogs. From its creeping stem rise branches that hold a pair of pink bells a half-inch long. Circumpolar in range, twinflower was named for the great Swedish botanist Linnaeus, who devised the modern binomial system of classifying plant and animal life.* (Jean T. Buermeyer)

188. *The scientific and common names for pipsissewa* (Chimaphila umbellata) *tell much about the plant and its lore. The Greek generic name means "lover of winter," for its foliage is evergreen and prominent in drab conifer woods across Eurasia and North America. The Latin specific name refers to its umbrella of waxy-white flowers. Pipsissewa is also a Cree Indian term meaning "juice breaks stone in bladder into small pieces." Pipsissewa leaves were once an important ingredient in root beer.* (Alvin E. Staffan)

190. *In classical mythology, Calypso was the queen of a Greek island who kept Ulysses concealed for seven years. Calypso, or fairy slipper (*Calypso bulbosa), *is also a lovely wild orchid, whose concealed habitat in boreal forests across North America, Europe, and Asia led botanists to give it such a romantic name. Wrote one orchid specialist in an unusual burst of poetry: "No other northern orchid has so captured the imagination of flower lovers as calypso! Its intricate and exotic beauty and early blooming day, the cool, mossy, enchanted quality of its environs, and its rarity near any populated area add up to romance and adventure for those who search for it."* (Kay McGregor)

191. *Drops of mist from Pacific fog bejewel a yellow violet (*Viola sp.) *along a trail in Olympic National Park in the state of Washington. Identification of violet species is often impossible even for botanists, for violets hybridize readily, and an individual plant might have several ancestors. Because their showy flowers appear in spring when rain or cold could prevent insect pollination, many violet species produce a second flower in summer; petal-less and closed, it is self-pollinating, to insure production of seed.* (Betty Randall)

192 *overleaf. The flowers of bunchberry (*Cornus canadensis), *a herbaceous member of the dogwood family of trees and shrubs, are among nature's many deceptions. The actual flowers, greenish-yellow and insignificant, are clustered in the center of four showy white, petal-like bracts. This late-spring carpet in a New Hampshire forest will produce tight clusters of scarlet berries in autumn. Found around the Gulf of St. Lawrence is a bunchberry with dark purple flowers; this is the very similar northern European species (*Cornus suecica), *the berries of which are used by Laplanders to make a tart pudding.* (Ed Cooper)

194. *Although snow lies deep across the valley, paintbrush* (Castilleja parviflora) *and lupine* (Lupinus latifolius) *are massed in a spectacular salute to autumn along Hurricane Ridge in Washington's Olympic National Park. There are some 200 species of paintbrush, found mostly in western North America. Its true flowers are inconspicuous: what actually catch the hiker's eye are the brightly colored bracts. The lupines likewise are a large and variable genus, one that tries the patience of professional botanists, let alone amateurs, who attempt to identify individual species. Perhaps it is enough simply to admire their beauty.* (Betty Randall)

A Flourish
on the Heights

Spring had come to mountainous Andorra. Although
the forested ridges and higher peaks of the Pyrenees
still had snow, the Soldeu Pass was open, permitting
us to get through it and see some of Andorra's
mountain wildflowers. At the crest of the pass, on the
edges of fast-disappearing snowbanks, spring was only
just beginning, with tiny pink primulas (*Primula*) awash
in meltwater. Hundreds of feet below the pass, we came
to streams bordered with yellow cowslips (*Caltha
palustris*). Here the adjacent mountain pastures were
in full bloom, colored with daffodils (*Narcissus
pseudonarcissus*), louseworts (*Pedicularis*), ground
orchids and the unbelievable blue of gentians (*Gentiana*).
Farther down in the valley, spring was already at its
peak, and countless poet's narcissus (*Narcissus poeticus*)
and other meadow flowers were triumphantly announc-
ing its climax.

Flowering scenes like these are duplicated in all the
mountains of temperate lands. The color of montane
wildflowers has a brilliance seemingly unmatched by
those of other habitats. Nowhere is this seen more
readily than in the flower fields of mountain meadows.
Such meadowlands are usually found near the timberline, that transitional belt between coniferous forest
and the treeless alpine zone with its open meadows,
talus slopes and rocky cliffs.

In many ways the alpine zone of temperate lands
resembles the tundra; it has some of the same or very
similar plants, and their growing season is also short.
The mountain wildflowers, which must complete their
life cycles as quickly as possible, seem to burst into

flower all at once. Many plants are retarded by the slow-melting snow, but some are so eager they even push their buds through the thinning snows to flower. Thus alpine meadows seem full of flowers throughout the short summer.

Above the timberline on tropical mountains grow some of the most unusual of alpine wildflowers. Good examples of these are found on the paramo, that cold, windy and wet alpine habitat of the tropical High Andes. Close to the equator day length scarcely changes, and daily temperatures at any given elevation are essentially constant. Within the alpine belt of the paramo the temperature on a rare bright sunny day may rise to 55°F (13°C), but generally it ranges in the 30s and 40s, with occasional frosts at night. Unlike the plants of arctic or alpine tundra, those of the paramo can maintain a constant, if slow, growth year-round.

Not far from the lofty Colombian city of Bogotá lies the paramo of Sumapaz. There I first became acquainted with paramo wildflowers, including the strange frailejón (*Epeletia*). Characteristically our botanical foray was greeted with mist, drizzle and cold, which at that 12,000-foot level penetrated our very bones. Yet we could still appreciate the paramo wildflowers.

The low thickets that are transitional from the lower woodlands to the higher paramo were full of flowering shrubs, primarily members of the daisy, heath and melastome families. Here we saw rose-pink *Bucquetia*, *Monochaetum*, red *Befaria*—the so-called rhododendrons of the Andes—mingled with Andean blueberries and huckleberries.

The shrubbery thinned as we climbed higher into the more characteristic soggy paramo grassland, inhabited by wiry bunchgrass and scattered bushlets with small leathery leaves. I recognized some of the herbaceous wildflowers as representatives of such familiar north temperate groups as buttercups, asters, violets and gentians. But with them were also New World tropical genera such as the spiked achupallas (*Puya*), characteristic terrestrial bromeliads of the High Andes, and especially the frailejones. Our first frailejones, with their thick cylindrical stems and terminal leafy rosettes, appeared in misty silhouette almost like cowled human forms. And so these curious endemic "rosette trees" must have appeared to the earliest Spanish travelers, for the Spanish *frailejón* means "friar-like." Later I saw thousands of them, like a religious procession marching across the El Angel paramo of northern Ecuador.

The silvery rosette of a young, still trunkless frailejón resembles a century plant (*Agave*) but has soft, woolly unarmed leaves. Unlike a century plant, however, which flowers but once and then dies, the frailejón produces branching clusters of attractive yellow daisy-like flowers almost continuously throughout its long life. Although some species remain simple stemless plants, most eventually produce the curious false trunk with its terminal leafy rosette. Some frailejones attain a height of thirty-five feet, but most reach only half that height. The dense whitish or silvery hairs that envelop all parts of the frailejón, forming a soft woolly mass, serve to reflect light. They are an evolutionary adaptation protecting the plant against excessive insolation ("sunburn").

Curiously, on other tropical mountains the same form of giant rosette herbs with silvery light-reflecting hairs has also been evolved—examples of evolutionary convergence. Thus on the high paramo-like grasslands of Kilimanjaro and its sister peaks in East Africa, one finds species of giant tree-like groundsels (*Senecio*) that look like the Andean frailejones and, to a lesser degree, so do the famed silverswords (*Argyroxiphium*) of Hawaii's high Haleakala peak. All three plants belong to the daisy family.

Frailejones are not found above 15,000 feet. Above that level the growing conditions for plants become progressively harsher as one approaches the snowline, between 16,000 and 17,000 feet. Snow and sleet storms replace the rains of lower paramo elevations, and frosts increase. A few large hairy rosette plants are still found, such as the Andean edelweiss (*Senecio*) and large lupines with giant flower spikes. But most of these high-paramo plants keep a low profile, compacting their leaves and stems and hugging the ground. Indeed the rosette plants are insignificant and often hidden in the shelter of tufted ichu grasses (*Stipa ichu*). Some, such as the dwarf composites, gentians and locoweeds (*Lupinus*), have large, colorful, but essentially stemless flowers, which seem to be growing right out of the ground.

At these higher elevations are found another paramo oddity, giant cushion plants. Representing several plant families, they occupy the bleaker sites; but unlike cushion plants of the tundra or of temperate mountains, those of the paramo grow continuously, if slowly, to produce huge cushions. Some of the strangest varieties exist on the puna of the lofty Peruvian altiplano. Puna is a kind of paramo that is dry and desert-like. On one trip across the puna I was fascinated by the large greenish boulder-like plants called yareta (*Azorella yareta*), of the carrot family. Most yareta cushions were low and spreading, forming great platters up to twelve feet in diameter. I found that I could walk on the plant's firm surface, apparently without harming it. Its cushion is a mosaic of tiny compacted leaf rosettes, dotted with insignificant greenish-white flowers. The seasonal aridity of the puna is one reason why many of its alpine wildflowers have adopted the cushion form. By crowding branches and leaves tightly together, they offer a minimum surface for water loss. The constant, near-freezing temperatures of the puna air also encourage the soil-hugging cushion form, for the plant takes advantage of a soil warmed by the equatorial sun. Stonefields, where yareta is most frequent, absorb even greater quantities of solar heat and, unlike more open ground, hold it for longer periods. Hence the attraction of yareta to rocky sites.

The "skyscraping" puna has some cacti, and these also form cushions. The well-known genus of prickly pears (*Opuntia*) is most commonly seen in the higher Andes and, like yareta, grows right up to the snowline. The larger cushion-forming types grow near bunchgrasses on the puna prairies. These areas, where llamas, alpacas and vicuñas graze, are subject to snow flurries, which may linger in scattered patches. At such times, cushion cacti and snow patches are almost indistinguishable, for the two commonest species, *Opuntia lagopus* and *Opuntia floccosa*, are covered with snow-white, felt-like hairs. These keep the inner part of the cushion nicely moist and protect it against prevalent drying winds.

Opuntia lagopus has cactus-like tiny yellow flowers; otherwise it is an aberrant member of its family, with branches closely packed into its woolly cushion. One can also stand on this firm plant mass, but beware, for *Opuntia* cushions have hidden spines and thus are more than ordinary plant cushions—they are virtual pincushions! Yareta and companion species of *Opuntia* represent the acme of the world's alpine cushion-forming plants.

199. *The appearance of the blue or lavender blossoms of pasqueflower* (Anemone patens) *signifies spring's arrival—if sometimes tentatively—on the mountain slopes and prairies of the American West. Wrote naturalist Ann Zwinger of a high meadow in Colorado: "When we come to the land in May, we go looking for the first pasqueflowers with dedication, hoping that we find them before the Drei Eisheiligen—the three icy sisters—brown the delicate blossoms. Though it is easy to be misled by the gleam of light on a silvery leaf or a tuft of snow, we find the first one in the first week of May. The emergent bud augers out of the ground, held tight within a furry spiraled involucre, often opening within the day." Most flowering plants send up leaves to produce and store food before they blossom; in contrast, alpine plants like the pasqueflower use carbohydrate reserves stored from the previous summer and unfurl their leaves after the flower has turned to seed. Furry hairs on the colorful sepals and stems trap the sun's warmth in these often cold days. The flower's name refers to its emergence about the time of Easter or Passover.* (C. A. Morgan)

200 *overleaf. The play of sun and shadow turns a wild iris* (Iris tenax) *on the slopes of Oregon's Mount Hood into an image even lovelier than the flower itself. Some 200 species of irises grow in northern climes, often in wet habitats. The genus bears the name of the Greek rainbow goddess because of the beauty and wide variety of colors found in its flowers.* (Gary Braasch)

203. *Both the common and generic names of cranesbill (*Geranium ibericum*) refer to the slender, pointed seedpod that develops from the flower's long pistil. To early botanists, it resembled the beak of a crane. Because its flowers grow upright, there is a tuft of hair at the base of each petal to protect the nectar from rain and dew. World-wide, there are some 400 species in the genus* Geranium. (Jane Burton/ Bruce Coleman, Ltd.)

202. *It is late August, but snow still lingers in an Austrian valley where small alpbells (*Soldanella pusilla*) swing in the wind. The petals of this charming flower of Europe's mountains are deeply cut to form a fringe. Its generic name, meaning "little soldo," refers to an Italian coin. (Eberhard Morell/ Roebild)*

204. *As many as a dozen lilac flowers grow on a single stem of this Rocky Mountain primrose* (Primula incana). *The name of this genus means "first," and primroses are indeed among the first flowers to appear when snow leaves mountain meadows between 8,000 and 12,000 feet. The higher ranges of Europe and Asia are especially rich in primroses.* (Larry West)

205 *top. On a boulder-strewn slope at 10,000 feet in Spain's Sierra Nevada range, a clump of purple violets* (Viola sp.) *soaks up the sun. Worldwide, there are some 300 species of violets, their flowers tinted red, purple, blue, green, yellow, cream, and white. Identification of species is often impossible even for botanists, for violets hybridize readily, and an individual plant might have several ancestors.* (Lennart Norström)

205 *center.* Flor de muerto, *or "flower of death," is the name Mexicans have given this dark gentian* (Lisianthus nigrescens), *for it is a favorite decoration at gravesites. Many flowers grow in a cluster atop its five-foot stem. The metallic sheen of gentian flowers is due to green-colored cells mixed with the dominant color of the petals.* (Steve Crouch)

205 *bottom. A prize discovery for hikers high in the Alps is the famous edelweiss* (Leontopodium alpinum), *subject of many songs and poems. The actual flowers appear as tiny yellow clusters forming a daisy-like head at the center of a star of fuzzy white, leafy bracts.* (Mary M. Thacher/Photo Researchers, Inc.)

206 *overleaf. Spectacular displays of avalanche lilies* (Erythronium montanum) *follow the melting snows in the Cascade and Olympic Mountains of Oregon, Washington, and British Columbia. Flowers in this genus are often called dog's-tooth violets because of their small, white pointed bulbs. For several years a bulb will produce only a single leaf, while patiently storing up enough food reserves to send up two leaves, a stalk, and a flower. Its generic name comes from the Greek word* erythros, *referring to the reddish-purple color of the Eurasian species.* (William A. Bake)

208. *On the lava slopes of Hawaii's volcanoes is found the remarkable silversword* (Argyroxiphium sandwichense), *which requires seven to twenty years of growth to reach the flowering stage. When this member of the daisy family attains a height of three to five feet, it produces upward of 400 reddish-purple flower heads and then dies. The silvery hairs on the leaves serve as mirrors to reflect the excessive sunshine at high elevations. Once, silverswords were so common they turned the cinder slopes of Mount Haleakala, on the island of Maui, silver-gray. Today they are rare, their numbers decimated by human exploitation early in this century and by introduced insects that attack the buds and the hungry feral goats that plague Hawaii's disappearing native flora.* (Murl Deusing/Photo Researchers, Inc.)

209. *The high mountains of East Africa—Kilimanjaro, Kenya, Meru, Ruwenzori, and others—are famous for their giant lobelias like this species* (Lobelia deckenii), *which on Mount Kilimanjaro grow in wet moorlands between 9,000 and 9,500 feet. Their extraordinary shaggy flower spikes may stand as high as twelve feet atop leafy stems of equal or greater height.* (H. von Meiss-Teuffen)

210 *overleaf. In the early morning light, the greens of sedges and the golds of groundsel* (Senecio sp.) *form a striking contrast in a mountain meadow in Oregon. Worldwide, there are some 1,300 species in this genus, one of the largest among flowering plants. The common name is derived from two Old English words meaning "pus absorber," for rural folk chopped up its leaves to treat abscesses.* (Gary Braasch)

212 *second overleaf. In the Arizona mountains, a shaft of sunlight spotlights the beautiful flower of golden columbine* (Aquilegia chrysantha). *The distinctive feature of columbines is the five spurs—one for each petal—which in this species of the American Southwest extend for nearly three inches. Columbines are pollinated by hummingbirds and insects capable of reaching the nectar at the tip of the floral spurs.* (Jack Dermid)

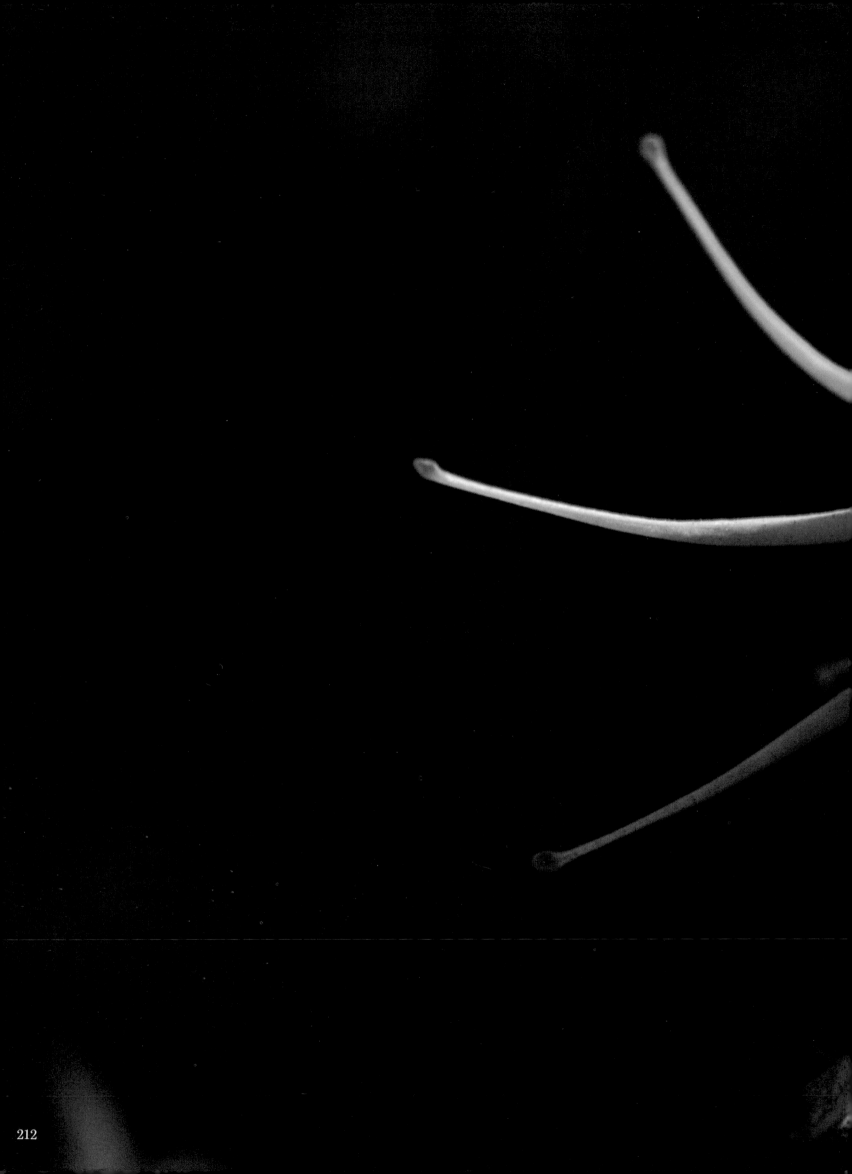

214. *The winged, deep pink, fragrant flowers of sweet vetch (*Hedysarum mackenzii*) sprawl over a stony stream bed in Alaska's Mount McKinley National Park. Sweet vetch is more common to Europe and Asia than to North America, but some species are found in the Rocky Mountains, where they are prized by mountain goats, high-dwelling rodents, and hikers. Mountain goats nibble the entire plant; the seedpods are stripped by marmots and pikas; and the nourishing roots, once gathered by Indians and trappers, have a licorice flavor. Sweet vetch belongs to the bean family, which botanist Harold William Rickett described as "one of the great families of plants; great in numbers, great in distribution over the Earth, great in importance to man." It includes, from Arctic to tropical climes, some 10,000 trees, shrubs, vines, and herbs, among them peanuts, soybeans, peas, clovers, and alfalfa. (Peter B. Kaplan)*

Survivors in a Cold World

Like giant snowflakes, countless cotton grasses
(*Eriophorum*) toss their silvery heads in the summer
sunshine in a wet arctic meadow. In a habitat not
noted for a great variety of plants, cotton grass is a
characteristic wildflower of the marshy plains of the
Arctic—barren and treeless, and first called tundra
by the Finns. In moist parts of the tundra, sedges
(cotton grass is actually a sedge, not a grass) and rushes
largely replace grasses. Their tufted tussocky growth
protects them against the desiccating action of wind.
But the wind, friend as well as foe, has widely distributed the light silky fruits, making cotton grasses
one of the most conspicuous of tundra wildflowers.
The treeless tundra with its permanently frozen
subsoil, called permafrost, covers vast inhospitable
areas of polar land. Frequently marshy, boggy or poorly
drained, it is as often rocky or stony, as in the extensive dry barrens, heathlands and fell-fields. Long, cold
and dark winters and short, cool summers with continuous daylight are among the climatic conditions
which most influence the growth and forms of arctic
wildflowers. Along with these are winter gales and
frequent summer frosts. Finally, scant rainfall makes
the tundra an unusual kind of desert.
Tundra is primarily a habitat of the far north, for the
mountainous Antarctic has scarcely any ice-free land
or wildflowers. Tundra thus forms a continuous but
uneven belt around the northern fringes and offshore
islands of Eurasia and North America. Its southern
boundary is the coniferous forest (taiga) where the
permafrost ends, permitting the establishment of trees.

The northernmost limit of tundra wildflowers is about 85° north latitude, roughly northern Greenland, where the region of permanent ice begins.

South of the Arctic, wherever mountains of the north temperate zone are high enough to produce a timberline, are isolated patches of so-called alpine tundra. Alpine tundra lacks important characteristics of true tundra, such as the twenty-four-hour days of summer. But among typical alpine wildflowers, such as moss campion (*Silene acaulis*) and numerous saxifrages, are many which migrated northward thousands of years ago. Thus a number of the wildflowers of both tundra and alpine tundra are shared by the taiga. The tundra climate is harsh and hostile for all vegetation. The relentless winds of winter fling abrasive particles of ice against all plants, whether in sheltered, snow-covered sites or forming prostrate mats, rosettes and cushions. Below ground the annual freezing and thawing action of permafrost makes the shallow soils unstable. The roots of flowering plants find life precarious in soil which is often pushed up into hummocks, slips down slopes, or is shaped into "frost boils" and odd polygonal patterns.

But the primary problem is winter temperatures that freeze a plant solid, and the low temperatures of the short summer, which hinder normal growth and reproductive processes. Cold also has an indirect effect, preventing absorption of water, a plant's lifeblood. Tundra water is largely frozen in permafrost or underground ice. In winter all water is unavailable. In summer the midnight sun, its rays weakened by its low-angled position in the southern sky, is seldom able to thaw more than a few inches of frozen soil. Plant root growth is thus severely restricted. Because water is so unavailable to plant life, the tundra has been called "the wettest desert in the world."

Tundra seasons are two—winter and summer. But summer has the short end, with scarcely two months for plant growth: mid-June to mid-August. Only then does the mean temperature stay above freezing. But the nearly continuous twenty-four-hour daylight is helpful. This is twice the total daily illumination, though not the warmth, received during the same period by plants in the tropics. Even in summer, temperatures in the Arctic may be low, but when the sun is out, the ground surface is always much warmer than the surrounding air. Arctic wildflowers have evolved special modifications to take advantage of this phenomenon. Many plants are prostrate or creeping, while other have compressed stems or leaves not only to restrict desiccation by reducing wind movement among leaves and flowers but also to aid, like the down of a parka, in trapping and holding warmer air around the plant body.

Flowers also serve as efficient collectors of solar heat. The darker the flower, the more heat it collects. However, the tiny tufted saxifrage (*Saxifraga cespitosa*), which has white or yellow flowers, offsets its color disadvantage by developing a deep red pigment in its leaves and shoots. On a sunny day in north Greenland the temperature within a saxifrage clump was recorded at 38°F (3°C) when the surrounding air was a cool 15°F (−9°C). Large single white flowers may also utilize flower form to gather warmth. The flower of the mountain avens (*Dryas integrifolia*) daily follows the path of the sun. Since the flower has a parabolic shape, it concentrates the solar heat on the central reproductive structures.

Summer begins like magic in the Arctic: one day marks the end of winter, the following the start of summer. Constant daylight permits tundra plants to flower and seed often with unbelievable rapidity. Snow buttercups (*Ranunculus*) complete their life cycle from flower to seed in a fortnight. Next year's growth must also be produced and packed in embryos in the over-wintering buds, ready to burst forth at first hint of summer. Since conditions in the arctic summer do not always favor seed production, the tiny creeping saxifrage (*Saxifraga flagellaris*) forms plantlets quickly at the ends of slender hair-like runners well before its seeds are produced.

The extended blossoming periods of temperate lands are telescoped into the first June weeks of the tundra summer. Among the earliest blooms, like their southerly relatives, are buttercups and cowslips (*Caltha*), bitter cress (*Cardamine*) saxifrages, willows and whitlow grass (*Draba*). Other wildflowers follow quickly. Arctic wildflowers may be few in species, but they often make up for this with an abundance of blossoms. The climate, so hostile to the vegetative parts of a plant, hardly seems to affect its flowers. A single plant of Arctic poppy (*Papaver radicatum*) may sport a hundred blossoms. The size of flowers may also appear far out of proportion to the size of the plant, for a single catkin of an Arctic willow (*Salix*) may be larger than the insignificant creeping shoot that gave it birth.

Although the Arctic landscape seems monotonous, there is variety in its habitats. Low-lying meadows and peaty depressions alternate with well-drained heathland or with the dry stony wastes of barrens and fell-fields. Each may have its own communities of wildflowers, which occur widely throughout the tundra. Thus the Arctic bell heather (*Cassiope tetragona*) and poppy, the Lapland rosebay (*Rhododendron lapponicum*) and the mountain avens range all round the tundra world.

The greatest variety of species is found at its southernmost borders, where the tundra merges with the northern coniferous forest. There in the "low Arctic," summer growing conditions are optimal, and vegetation of some sort usually covers the ground. In moist areas tundra shrubs attain maximum size, and perennial wildflowers may form large, showy colonies. The dominant cotton grasses share such meadowlands with fireweed, daisies, knotweed, lupines and louseworts, yellow marsh saxifrage (*Saxifraga hirculus*), and various orchids. On sheltered slopes, especially where banks of snow accumulate in winter, are thickly matted swards dominated by Arctic heaths.

The rocky barrens and fell-fields possess the driest, least hospitable of the tundra habitats. But they too have their typical plants such as mountain avens and the insignificant rosette or cushion-forming herbs. All seek shelter, often in south-facing depressions among boulders and rocks. When in flower, these oases may produce tiny rock gardens vivid with color.

As we near the Pole, growing conditions rapidly degenerate. Vegetation becomes sparse, and individual plants are more widely scattered and limited to sheltered spots. A point is reached in the "high Arctic" —at about 85° north latitude—beyond which wildflowers cannot grow; beyond is only a desolate, icy desert landscape. At that high latitude—about 400 miles from the Pole—the feeble slanted rays of the summer sun are unable to melt even the surface of the frozen ground. All higher plant life ceases. Reaching the fringes of this "no-plants-land" is a little band of flowering plants that includes mountain avens, purple saxifrage (*Saxifraga oppositifolia*), Arctic poppy, crowberry and Arctic bell heather. Certainly these must be classed as the world's hardiest wildflowers.

219. *Twilight at 11,000 feet on Colorado's Mount Evans intensifies the lavender of moss campion (Silene acaulis), a circumpolar species that belongs to the far north but reaches southward along the spines of high mountain ranges. The leaves of moss campion form a cushion an inch high and sometimes a foot wide, its dense foliage a defense against chilling winds. Likewise, the unusual abundance of red-pigmented flowers—its quarter-inch blossoms are most often pink, rarely white—collects solar heat. On alpine meadows and stony ridges in the Rocky Mountains, moss campion blooms in early July when, as one naturalist wrote, "the water pipit, white-crowned sparrow, and gray-crowned rosy finch are laying eggs." In the high Arctic, where landmarks are few, moss campion is a reliable compass, for its flowers are grouped on the south-facing side of the cushion where light intensity is greatest. (Larry West)*

220. *An early August snowfall on the north slope of Alaska's Brooks Range fails to deter the progress of monkshood* (Aconitum delphinifolium), *its helmet-shaped flowers formed by purple sepals. All parts of the monkshood plant are poisonous, and this is true of the several hundred other species in this genus of the buttercup family. From one* Aconitum *in Nepal is extracted a deadly poison called* bikh, *used in many murders over the centuries. Yet the toxic alkaloids in these plants also are used to make a drug called aconite, a heart and nerve sedative. The curious monkshood flower—its petals transformed into nectar glands—is designed for bumblebee pollination, but these insects often avoid the fancy internal mechanism by biting right through the hood.* (Les Line)

221. *One must kneel to appreciate the beauty of many tundra plants, and none more so than mountain forget-me-not* (Eritrichium aretoides), *whose minute sky-blue flowers stand less than an inch above the ground. Of another high-country forget-me-not, the biologists John and Frank Craighead wrote: "Perhaps no flower gives the mountain lover more joy than this little blue gem. . . . Its beauty cheers the weary climber, and its presence tells him that he is nearing the summit. It forces him to stop and consider how such delicate beauty survives the cold and storms of the mountains, why such a lovely thing is hidden from the eyes of most men."* (Stephen J. Krasemann)

222 *overleaf. To reproduce, this tundra saxifrage* (Saxifraga flagellaris) *extends long, naked runners from its rosette of basal leaves. New plants grow from the tips when they embed themselves in the soil. The specific name means "whip-like."* (Fred Bruemmer)

225. *Of the thirty kinds of shooting star native to western North America, only one species reaches the Alaskan tundra. Alpine shooting star* (Dodecatheon frigidum), *its specific name clearly indicating the plant's cold haunts, also is found across the Bering Strait in Russian Siberia. In emergencies, the roots and leaves can be cooked as survival food.* (Erwin A. Bauer)

224. *An early foraging butterfly has been attracted by the bedewed golden rays of alpine arnica* (Arnica alpina) *in an Alaskan valley. A tincture made from the aromatic flower heads of various species of arnica was once widely used to treat bruises and sprains.* (Robert Belous/Alaska Task Force)

226 *overleaf. Plants in the genus* Oxytropis *sometimes are called crazyweed or locoweed because they cause fatal spasms in grazing animals. Of some 300 species, a few, such as* Oxytropis nigrescens, *reach the Arctic tundra. The pointed keel of the flower identifies it as a member of the bean family.* (Stephen J. Krasemann)

228. *Although forget-me-not is the state flower of Alaska, the plant most typical of that Arctic state's vast expanses of boggy tundra is the sedge known as cotton grass (Eriophorum sp.). Its silky floral envelope dances in the winds that blow off the Bering Sea or Arctic Ocean.* (George W. Calef)

229 *top. The fragile blossoms of northern windflower (Anemone parviflora) are transparent in the midnight sun of the northern Yukon Territory. A hundred species, mostly inhabitants of woods and meadows, belong to this genus; its name is derived from the Greek for wind, anemos. Several stories explain this connection: according to one legend, the flowers open at the command of the spring breeze. Tiny flowers on delicate stems, they certainly tremble at the slightest gust.* (George W. Calef)

229 *below. Of all tundra flowers, the best-known is the Arctic poppy (Papaver radicatum), found around the Pole. This poppy, which prefers dry open habitats, is so variable that botanists cannot agree on its classification, and a number of species and subspecies have been described by the "splitters" in taxonomic ranks. Like other tundra plants, the leaves of Arctic poppy are humble—but as many as a hundred flowers appear on a single plant.* (Fred Bruemmer)

230 *overleaf. Eider ducks, purple sandpipers, and dovekies were nesting on an island in northern Spitsbergen when this clump of snow saxifrage (Saxifraga nivalis) burst into flower. This species is typical of tundra representatives of its genus, for its leaves are semi-evergreen. They are produced one summer, remain green through the winter, and give the plant a prompt start the next growing season. The name saxifrage is derived from the Latin meaning "to break stones," an allusion to the rocky habitat among which many species live.* (Arthur Christiansen)

232 *second overleaf. The tiny bells of moss plant (Cassiope hypnoides), perhaps a fifth of an inch long, bob in the wind that rakes the summit of New Hampshire's Mount Washington. Cassiopes are low evergreen heaths whose apparent delicacy belies their capacity to survive in desolate and bitter habitats—lands above the Arctic Circle and high mountaintops farther south. Without their flowers, they do indeed suggest clumps of moss; one species in Alaska and Siberia is named after the club moss, Lycopodium.* (Jean T. Buermeyer)

234. *Dozens of tiny dandelion fruits, each rigged with its little parachute, await distribution by the wind—or by a small child puffing on the "blowball." Common dandelion* (Taraxacum officinale) *is a ubiquitous, persistent, pestiferous weed that is nearly impossible to eradicate: its flat leaves evade lawnmowers; its tough, deep roots resist yanking; and it produces several generations a year. Dandelions are a particular bane to fruit growers, for their butter-yellow flowers lure bees away from apple blossoms in need of pollination. Yet no one can deny the beauty, if commonplace, of a vast field of dandelions. Moreover, they are a gourmet's delight: their young leaves are an excellent salad green or potherb; their buds can be boiled or pickled; their flowers, dipped in batter and fried, are a treat; their roots, baked and ground, make a delicious coffee substitute; and, perhaps best of all, their flowers are the key ingredient in a splendid, golden wine tasting of summertime and the country.* (Ingmar Holmåsen)

A Common Beauty

Ours is a rural home in central New York state, and the adjacent fields and roadsides serve us well as a country florist. The first daisy gives as much pleasure as a garden rose. From June until frost we have a varied offering—at first buttercups and hawkweeds and then a variety of clovers. Early summer brings the lacy wild carrot (*Daucus carota*) and St.-John's-wort (*Hypericum perfoliatum*), followed later by lotus, teasels (*Dipsacus*), and purple loosestrife (*Lythrum salicaria*). Some of our flowers—the day lily (*Hemerocallis*), sky-blue chicory (*Cichorium intybus*) and purple vetch (*Vicia*)— are not suitable for cutting, nor are the coarser mullein (*Verbascum*), burdock (*Arctium*) and thistle (*Cirsium*), but all bring pleasure.

All these "wildflowers" are immigrant weeds; like most Americans, their forebears are of Old World origin. Sharing a similar ancestry are other less showy weeds that we constantly fight in our gardens — chickweed (*Stellaria*), dock (*Rumex*), mustard (*Brassica*), purslane (*Portulaca*), plantains (*Plantago*) and pesky grasses. Over a thousand kinds of weed flowers—an amazing one-fifth of the total flora— thrive in northeastern North America alone!

How did these Eurasian plants get to America? All were carried here by man. Day lilies, loosestrife and bouncing bet (*Saponaria officinalis*) were brought in for the flower garden, whereas clovers, trefoils and the valuable meadow grasses came as much-appreciated forages. And since agricultural seed is seldom pure, the seeds of many common weeds of European pastures, such as the oxeye daisy (*Chrysanthemum leucan-*

themum), bindweeds (*Convolvulus*) and chicory, were introduced with the forages. And most thrived.

Once settled in the new land, species aggressive enough to compete with native plants "escaped" and became a part of the local flora. Thus weeds, as well as cultivated plants, have followed man around the world. For as John Burroughs said: "Weeds are great travelers; they are, indeed, the tramps of the vegetable world. . . . They walk; they fly; they swim; they steal a ride; they travel by rail, by flood, by wind; they go underground, and they go above, across lots, and by the highway: in the fields they are intercepted and cut off; but on the public road, every boy, every passing herd of sheep or cows, gives them a lift."

What is a weed? "Any undesired, uncultivated plant." So says the dictionary, while an English gardener affirms that "a weed is but a good plant in the wrong place," adding: "daisies [that is, English daisies] are not perhaps in their right place in lawns but I should be sorry to see my lawn quite free from them." Many feel the same way about dandelions (*Taraxacum*).

Weeds are aggressive invaders. Mostly sun-lovers, they occupy land where the natural vegetation has been disturbed, often by man. Forests and other stable habitats seldom attract weeds, for there the competition of other plants is usually too great. But in the north temperate zone fireweed appears like magic, temporarily covering areas burnt over by forest fires; or goldenrods (*Solidago*) assume command of abandoned farmland; while roadsides sport a varied crop of weedy wildflowers.

Weeds are always waiting in the wings to take over nature's stage. The wildflowers of London's blitz are good examples. Weeds moved in like magic to convert that city's rubble to colorful gardens after the air raids in the early 1940s. Most common "blitz flowers" were fireweed (*Epilobium*), horseweed (*Erigeron*) and ragwort (*Senecio*), but botanists found ninety-three other invading species on the bombed sites. The trio of weeds that dominated the scene did so because of the efficiency of their seed dispersal; all three are wind-distributed, using the same kind of parachute mechanism seen in such other familiar weeds as the dandelion, milkweed, goldenrod and thistle.

Other wildflowers become invasive weeds primarily through efficiency in vegetative rather than seed reproduction. Thus a single plant of hawkweed or runaway-robin (*Nepeta hederacea*) sends out creeping stems in all directions, multiplying its numbers phenomenally in a single season without seed production.

Every land has its own native weeds. Because such plants evolve along with their own natural ecological checks and balances, they often are inclined to be "shy and harmless." But free such plants from their biological controls in a foreign environment, and they show a different character. Thus, in its native Europe, the dandelion seldom shows the weedy abandon it displays as an immigrant abroad.

As is evident, civilization with its penchant for upsetting nature's system of biological checks and balances has encouraged the spread of weeds. The water hyacinths (*Eichhornia crassipes*) of tropical America, among the most attractive of aquatic wildflowers, are masters of vegetative reproduction. Within a few weeks a clump of hyacinths can double its number. Thus these hyacinths, escapees from cultivation, now clog the lakes and waterways of tropical Africa and Asia and in the Deep South of the United States in frightening numbers. Today it costs millions of dollars annually to clear important waterways of this pestiferous plant.

Because of its rapid vegetative reproduction, one might have guessed that the water hyacinth could become a potent weed. This is hardly the case with the Japanese honeysuckle (*Lonicera japonica*). In Nippon this attractive flowering vine is in balance with other plants in its deciduous forest habitat. A century ago the species was introduced as a prospective new ornamental for the gardens of the northeastern United States, and it became a popular garden subject there: the cold winters kept its unsuspected aggressiveness under control. But when transplanted to the southeastern United States, where growing conditions are ideal, the Japanese honeysuckle became a terrible pest. It soon escaped to lowland forests, where it can be seen smothering the native undergrowth, including young trees needed to perpetuate reproduction of the deciduous forest.

Wherever alien wildflowers appear as successful weeds on a local scene, they usually have come, without their natural biological controls, from an identical environment elsewhere. Just as the field weeds of temperate Europe have become even more aggressive weeds as immigrants in the meadows of other temperate lands, so regions with a Mediterranean type of climate usually find their weeds are immigrant wildflowers from sister climates elsewhere in the world. Thus in Mediterranean Australia the naturalized Capeweed (*Arctotheca calendula*) and various weedy species of oxalis are of South African origin, while in South Africa one finds that certain *Hakea* of Australia are a serious local pest.

Nowhere is the intercontinental exchange of plants more apparent than in the wet tropics. Hundreds of weedy wildflowers of the paleotropics prove to have been originally natives of the New World tropics, and vice versa. Thus the weeds of the West Indies are natives of the East Indies, and those of the East Indies came from the West. It is interesting that many of these hitchhiked on the Spanish galleons that once plied the trans-Pacific trade route between Acapulco and Manila via Guam. The greatest concentrations of spontaneous American weeds appeared initially in Guam and the Philippines, while Mexico was the gateway for immigrant weeds from tropical Asia. Today tropical weedy wildflowers like the little blue ageratum (*Ageratum conyzoides*) and pink sensitive plant (*Mimosa pudica*), originally from the New World, are essentially pantropical.

Just how domineering immigrant weeds can become is shown in the Hawaiian Islands. That archipelago was originally forested and had an almost completely endemic native flora. The early Polynesian settlers brought some new plants to the islands, but what really transformed the Hawaiian vegetational scene were Europeans who introduced many new plants with their advanced agriculture. After they cleared the land, which was accompanied by aggressive colonization by immigrant plants, the native vegetation with its unique wildflowers was either seriously disturbed or completely destroyed. Today visitors to Hawaii—and even her residents—find it difficult to locate a truly native wildflower. Even the hibiscus (*Hibiscus rosa-sinensis*), the state flower, is an alien from Asia.

How easy, then, to agree with the American naturalist John Burroughs, who wrote that "the most human plants are the weeds. How they cling to man and follow him around the world. . . . How they crowd around his barns and dwellings, and throng his garden and jostle and override each other in their strife to be near him! . . . one comes to regard them with positive affection."

239. *This scene could be found on either side of the Atlantic Ocean—a field utterly choked with two troublesome, though attractive, weeds that were native to Europe and carried to North America by early colonists. The yellow flowers belong to tall buttercup (Ranunculus acris), which carpets English meadows with gold but is a bane of dairy farmers, for cows grazing on this poisonous plant give unpalatable milk. The white flowers belong to a wild morning glory, field bindweed (Convolvulus arvensis), known in England as hellweed. The Latin name of this genus means "to entwine" and refers to the plant's habit of twisting tightly around the host plant on which it climbs. Its funnel-shaped flowers, often pink, open before dawn and wither shortly after noon. (Tom Myers)*

240 *overleaf. Queen Anne's lace (Daucus carota) is the original carrot; through 2,000 years of cultivation, the stringy white taproot of the wild plant has become the fleshy orange vegetable so familiar on our tables. Before its flower cluster is in full bloom, and later when its fruits are ripening, the umbel bends inward in the shape of a bird's nest. The tiny fruits are hooked to ensure their dispersal in the fur of passing animals. (Sonja Bullaty)*

242 *second overleaf. Introduced to North American gardens from Europe, purple loosestrife (Lythrum salicaria) became an aggressive invader of wetlands and floodplains, crowding out native plants that are important food supplies for hard-pressed waterfowl. Its single saving grace is that a swampy meadow crammed with four-foot spikes of its sweet-scented flowers is indeed a breathtaking sight. (Susan Rayfield)*

244 *third overleaf. In ancient times it was believed that birds of prey owed their incredible eyesight to eating the flowers of hawkweed, and herbal doctors made eye lotions from the juice of these widespread plants. Orange hawkweed (Hieracium aurantiacum), a European native that overran American fields and pastures, was damned by farmers as "devil's paintbrush." (Susan Rayfield)*

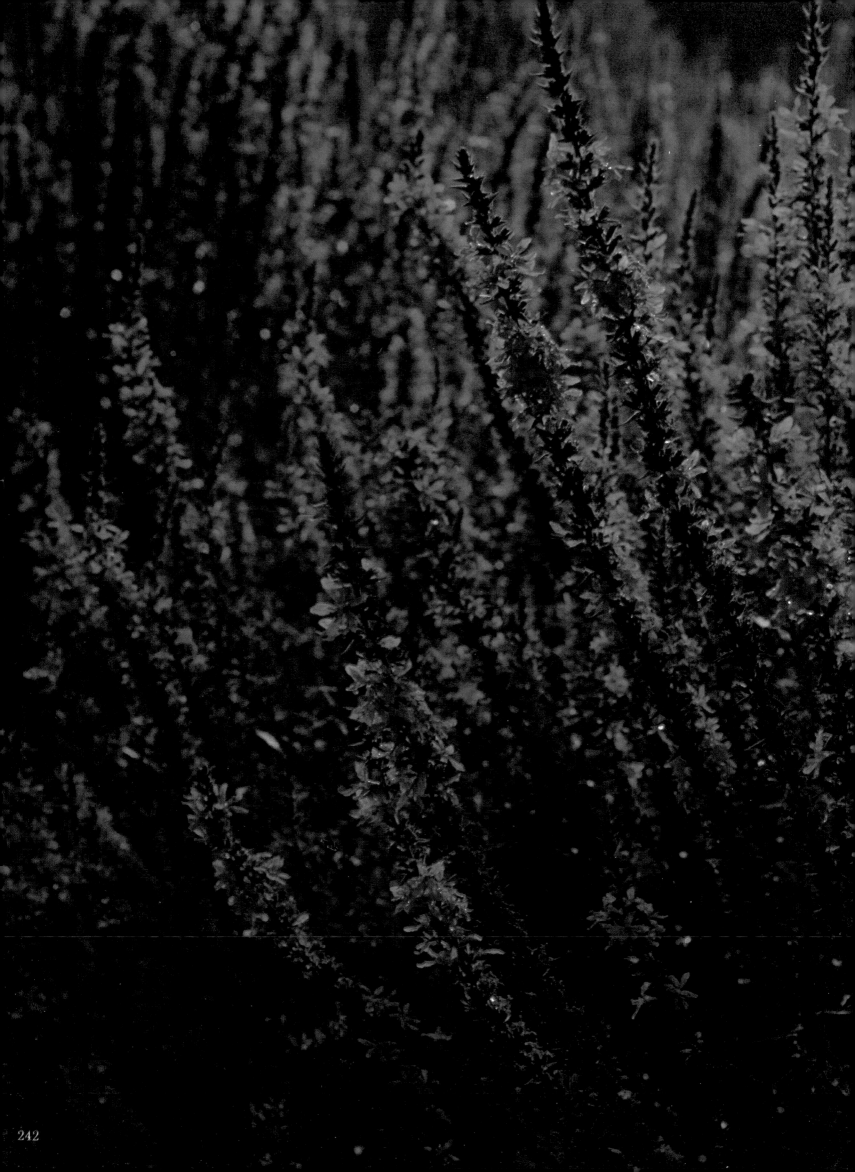

246. *Kin of clovers and alfalfa, bird's-foot trefoil (Lotus corniculatus) was an important forage plant in Europe and hence was carried to faraway lands by colonists. In North America it became a common road-side weed. The "bird's foot" is suggested by the shape of its leaves.* (Hans Reinhard/Bruce Coleman, Ltd.)

247. *A single flower of scarlet milkweed (Asclepias curassavia), one of perhaps fifty in a cluster, shows its extraordinary structure. The five curved-back red petals support a tube formed by the stamens. From this rise five golden, nectar-filled cups that attract bees, butterflies, and moths. Between each cup is a slit that can trap the leg of a feeding insect; when it struggles free, it carries tiny sacs of pollen to the next flower it visits. The "milk" of milkweed is a white latex that runs from a broken stem or leaf; it contains toxins that a few insects, such as the famous monarch butterfly, are able to store as a defense against predators.* (James H. Carmichael, Jr.)

248 *and* **249.** *The crinkled new petals of Europe's common poppy* (Papaver rhoeas) *surround the pistil and a fringe of stamens tipped with black pollen. The four paper-thin scarlet petals drop off soon after they unfurl, leaving the pistil to mature into a capsule filled with tiny black seeds. Another European poppy, with flowers of lilac or white, is the source of seeds used to add zest to breads and cakes and to yield oil used both for salads and in the manufacture of paints. This is the notorious opium poppy, whose unripe capsule, if scratched, oozes the raw narcotic from which heroin, morphine, and codeine are derived. Poppy seeds imported into the United States must first be sterilized to prevent the flower from being grown here.* (Jane Burton/Bruce Coleman, Ltd.; Pierre A. Pittet)

250 *overleaf. Few North American weeds are more familiar than oxeye daisy* (Chrysanthemum leucanthemum), *originally a native of Europe and Asia. Its white rays surround a mass of tiny flowers; those in the central whorl are the fertile reproductive flowers, while the sterile outer ray flowers attract insects. Worldwide, there are hundreds of species of wild chrysanthemums. The dried flower heads of several kinds yield pyrethrum, an insecticide that is harmless to other animal life.* (Michael and Barbara Reed/ Earth Scenes)

Notes on Photographers

Ruth Allen (75, 151) is a portrait artist turned nature photographer. She studied taxonomy at the University of Pennsylvania. Her photographs appear in each of the many volumes of Rickett's *Wildflowers of the United States.*

William A. Bake (76, 86, 158, 206-207) is Artist-in-Residence at Appalachian State University. His photographs and writings have appeared in numerous magazines including *Reader's Digest* and *Popular Photography* and such books as *The Blue Ridge.*

Anthony Bannister (62) is a professional photographer living in England.

Erwin A. Bauer (14, 26-27, 67, 225) is a freelance writer-photographer who specializes in adventure and wildlife. His most recent book is *Hunting with a Camera,* a world guide to wildlife photography.

Robert Belous (224) is a well-known nature photographer who lives in Jackson, Wyoming. His work has appeared in *Audubon* magazine and other publications.

Gary Braasch (147, 200-201, 210-211), a freelance photographer living in Vancouver, Washington, is director of the Northwest Environmental Defense Center. His nature photographs have appeared in numerous magazines, calendars and Time-Life books.

Stanley Breeden (29, 51, 95, 96) is an Australian photographer whose work has appeared in *National Geographic, Audubon* and many books.

Fred Bruemmer (222-223, 229) has photographed wildlife all over the world. He is the author of *The Arctic* and his work has appeared in many international publications, such as *The Audubon Society Book of Wild Animals.*

Jean T. Buermeyer (189, 232-233) has been photographing wildflowers since 1967. A graduate of Tuft's University, she frequently lectures on the wildflowers of New England.

Sonja Bullaty (167, 168-169, 173, 240-241) has published photo essays in *Audubon, Horizon* and *Life* magazines. Winner of numerous awards, her work has also appeared in many books, such as *The Wild Places.*

Jane Burton (203, 248) is a photographer-artist who lives with her husband in Surrey, England in a converted school house that serves as both her home and studio.

George W. Calef (228, 229) is a wildlife biologist with the Fish and Wildlife Service of Canada in the Northwest Territories. He was recently the leader of a project to study the effect of the Arctic Island pipeline on Canadian wildlife.

Bob and Clara Calhoun (67) are a husband and wife team of naturalist-photographers. They have been members of the Photographic Society of America for the past 16 years and their work has appeared in numerous publications including *National Wildlife, Natural History* and the *Audubon Society Field Guide to North American Birds.*

James H. Carmichael, Jr. (22, 23, 32, 36-37, 44-45, 50, 68-69, 102-103, 247) is a professional photographer who also teaches nature photography in Sarasota, Florida. His work appears in numerous magazines and books.

Patricia Caulfield (106), a professional photographer living in Las Vegas, is a frequent contributor to many publications, including *National Geographic.* She is the author of *Everglades,* published in 1970.

Eric Chricton (97) is a British freelance photographer who specializes in natural history. He has traveled extensively in Europe, Australia and New Zealand and his work appears in numerous books, magazines and encyclopedias.

Arthur Christiansen (230-231) lives in Copenhagen and has been photographing wildlife since 1928. He has conducted expeditions to Greenland, Iceland, Spitzbergen, most of Europe and the Mediterranean, and his work has appeared in hundreds of books and magazines around the world.

Ed Cooper (10-11, 70, 182, 192-193) lives in Everett, Washington and has captured the images of America from Maine to Maui. A professional photographer since 1968, his work has appeared in numerous publications, including the cover of the July, 1977 issue of *National Geographic*.

Steve Crouch (78, 79, 80-81, 205) has been a professional photographer since 1948, specializing in editorial and advertising photography. His two books, *Steinbeck Country* and *Desert Country*, have been circulated widely.

James Cunningham (176-177) is a photographer from Connecticut currently living in Grenados.

Jack Dermid (132, 212-213), a widely known nature photographer, lives in Wilmington, North Carolina. He is co-author of *The World of the Wood Duck*, and has completed a motion picture short entitled *The Living Coast*.

Murl Deusing (208) is a photographer for the Walt Disney True-Life Adventure series, and the producer and host for the Murl Deusing Safari televison series. He is a nationally known lecturer and a producer of classroom films.

Jon Farrar (87) has contributed to *Audubon*, *Natural History*, *Outdoor Life*, and *The Audubon Society Book of Wild Animals*. His special interest is the plants and animals of the North American grasslands.

M. P. L. Fogden (20-21, 118-119, 130-131) is a professional biologist with a doctorate in ornithology from Oxford University. He has spent the past 15 years researching wildlife in Borneo, Uganda and Mexico. His photographs have appeared in many books and magazines and he is co-author (with his wife) of *Animals and Their Colours*.

François Gohier (46) is a French wildlife photographer whose work has been published in magazines and books throughout the world. His photographs most recently appeared in *Wildlife of the Mountains*.

Patrick W. Grace (178) is a freelance photographer living on the coast of Maine who specializes in close-up nature photography. His work has appeared in numerous magazines, newspapers and textbooks, including *Natural History* and the *New York Times*.

Farrell Grehan (54, 65, 90, 104-105, 174) is an internationally known photographer whose work has appeared in magazines and books worldwide. Formerly a *Life* photographer, he is currently on assignment for *National Geographic*.

Walter H. Hodge (30, 31, 47, 48, 49, 51, 98-99, 100, 111, 114-115, 127) the co-author of this volume, is a leading American botanist. A photographer of natural history throughout the world, he has authored nearly 200 papers on botany, horticulture and general natural history.

Ingmar Holmåsen (116, 172, 179, 234) has been a professional nature photographer in Sweden since 1966. Educated at the Grafiska Institute in Stockholm, he is the author of six photo-illustrated books, one of them on nature photography.

George Holton (122) is a New York-based photographer specializing in natural history, primitive peoples and ancient architecture. His work appears in many publications, including *Audubon* and *National Geographic*.

M. Philip Kahl's (19) work has been reproduced in many magazines and books, including *National Geographic* and *The Audubon Society Book of Wild Animals*. He is a naturalist who specializes in the study of storks and flamingos of the world.

Peter Kaplan (214) specializes in nature and wildlife photography. His photographs have appeared in *Audubon*, *Natural History* and other magazines as well as *The Audubon Society Book of Wild Birds*.

Stephen J. Krasemann (64, 83, 221, 226-227) is a Wisconsin-based nature photographer whose work has appeared in *Audubon*, *National Geographic*, *National Wildlife* and many other magazines. He has contributed to many books on natural history, including *The Wild Shores of North America*.

Dwight R. Kuhn (149) is a Maine photographer and high school teacher specializing in the natural sciences. His photographs have appeared in books published by The National Geographic Society, the National Wildlife Federation, and in numerous magazines.

Carl Kurtz (1, 2-3, 4-5, 77), a naturalist and farmer living in central Iowa, specializes in prairie botany and grassland restoration. His work has appeared in *Audubon, National Wildlife* and many other magazines and books.

Les Line (170-171, 220), co-author of this book, was born in Michigan, and won many prizes for news photography and conservation reporting before becoming editor of *Audubon* in 1966. The magazine has since be-become known for its excellent graphics, color photography and environmental reporting.

Kay McGregor (190) is an amateur photographer with a particular interest in wildflowers. She is an Associate of the Photographic Society of America and the Royal Photographic Society of Great Britain.

Wendell D. Metzen (140-141, 154), a biologist-photographer living in Georgia, has been published in *National Geographic, National Wildlife,* Time-Life books and Sierra Club calendars.

Robert W. Mitchell (156, 157, 179, 180-181) is a freelance photographer and associate of the Department of Biology at Texas Tech.

Michael Morcombe (100, 101, 142) is an Australian naturalist, photographer and writer specializing in the Australian bush. He is the author of *Wild Australia* and *Birds of Australia* and his work has appeared in *The Audubon Society Book of Wild Animals* and other publications.

Eberhard Morell (202) lives in Frankfurt, West Germany and has been photographing nature since the age of 14.

C. Allan Morgan (66, 67, 199) has been a freelance wildlife and natural history photographer for ten years with photographs appearing in many national publications. Since 1975 he has been a photographer for the Arizona-Sonora Desert Museum near Tucson, Arizona.

David Muench (60-61) is a Californian who specializes in photographing the western landscape. Among his many book credits are *California* (with Ray Atkeson), *Arizona* and the forthcoming *Desert Images: An American Landscape* (with Edward Abbey).

Ernst Müller (152-153) is a professional animal photographer living in Frankfurt, West Germany.

Sally Myers (6-7) has been a freelance photographer for 15 years with her husband Tom. Her photographs have appeared in numerous books, newspapers, educational film strips and magazines such as *National Geographic.*

Tom Myers (239), a natural history photographer living in Sacramento, California, has been published in *National Geographic* and in many books, including *The Wild Shores of North America.*

Lennart Norström (205) is a nature photographer living in Sweden.

Oxford Scientific Films (117, 148, 150, 155), started in 1968 by seven professional biologists from Oxford University, makes natural history and educational films and takes stills for use as book illustrations and in educational film strips.

Carroll W. Perkins (88-89, 128-129, 178, 187) is a chemist who prefers to be known as a photographer. His work appears regularly in Canadian nature publications such as *Nature Canada* and the *Ontario Naturalist.* His work has also appeared in books by Reader's Digest and in *Butterflies of the World.*

Pierre A. Pittet (249), living in Switzerland, is a photographer of natural history whose work has appeared in numerous publications.

D. C. H. Plowes (62, 63) lives in Rhodesia and specializes in photographs of natural history subjects. His photographs have appeared in numerous publications throughout the world. The plant *Aloe plowessi,* discovered by him in 1950, is named in his honor.

Allan Power (51) lives in New Hebrides where he photographs all aspects of that island's life. His work has appeared in numerous publications.

Betty Randall (59, 149, 191, 194) received a M.A. in biochemistry from the University of Southern California at Berkeley. She has spent 20 years as a laboratory researcher and her photographic work has appeared in Sunset books, *The Wild Shores of North America* and *Audubon* magazine.

Susan Rayfield (120-121, 136, 242-243, 244-245), formerly picture editor at *Audubon* magazine, is now the senior editor at Chanticleer Press. Her work has appeared in *The Wild Shores of North America*, Reader's Digest and Time-Life books, *Audubon* and many other publications.

Michael and Barbara Reed (250-251) specialize in anthropological photography. Their work has taken them to Central America, Africa, the Galápagos and the West Indies.

Hans Reinhard (166, 246) lives on a small farm outside Heidelberg with his family and a menagerie of wild animals. He is the author of *Die Technik der Wildphotografie*.

Dorothy M. Richards (179) is a freelance nature photographer living in Dayton, Ohio.

Edward S. Ross (24-25, 51, 84-85, 100, 101), Curator of Insects at the California Academy of Sciences, pioneered in candid insect photography in his *Insects Close Up* (1953). He has photographed nature in almost every part of the globe and his work appears in hundreds of publications.

Kjell B. Sandved (33, 34, 35, 63) produces biological motion pictures in conjunction with scientists at the Museum of Natural History of the Smithsonian Institution. He also lectures throughout the country for the Smithsonian Associates on the subject of animal behavior.

John Shaw (43, 137, 138, 139, 150, 163, 164-165, 178) is a professional nature photographer living in Harrison, Michigan. His photographs have appeared in magazines such as *National Wildlife*, *Audubon*, *International Wildlife* and in numerous books, film strips, textbooks and advertisements.

Alvin E. Staffan (82, 112-113, 188) is both a nature photographer and an artist for the Ohio Department of Natural Resources. His photographs appear in many books, magazines and encyclopedias. He lectures frequently on nature photography.

Mary M. Thacher (205), the wife of a U.N. official on assignment in Kenya, is a photographer and artist with a degree in printmaking from the Ecole genevoise des Arts Visuels in Geneva, Switzerland. Her photographs appear in numerous books for children, textbooks and encyclopedias.

Katrina Thomas (28) is a New York-based photographer whose specialty is people. She has photographed two books for children about New York City, *Chiro* and *My Skyscraper City*, and has done reportage photography for numerous organizations including the National Park Service.

Michael Tweedie (52-53) has worked for the past 15 years as Consultant Zoologist for *Wildlife* magazine. Director of the Raffles Museum in Singapore from 1946 to 1957, he has published numerous papers and several books on Malaysian zoology.

H. von Meiss-Teuffen (209) is a German-Swiss photographer who travels extensively and has been a professional photographer for over thirty years. His work appears in many books and periodicals internationally.

Karl Weidmann (38), a freelance photographer and producer of nature films, lives in Venezuela. His work has appeared frequently in publications around the world, including *National Geographic*.

Larry West (133, 134, 135, 175, 219) is a professional nature photographer living in Mason, Michigan. His photographs appear in many magazines and books, including *The Wild Places* and *The Audubon Society Book of Wild Birds*.

Index

Numbers in italics indicate pictures